A practical approach to
toxicological investigations

A practical approach to toxicological investigations

A. POOLE

*Assistant Director of Toxicology, Department of Toxicology,
Smith Kline and French Research Ltd, Welwyn, UK*

G. B. LESLIE

Consultant Toxicologist, Bioassay Ltd, Biggleswade, UK

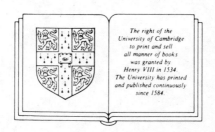

The right of the
University of Cambridge
to print and sell
all manner of books
was granted by
Henry VIII in 1534.
The University has printed
and published continuously
since 1584.

CAMBRIDGE UNIVERSITY PRESS

Cambridge
New York   Port Chester
Melbourne   Sydney

Published by the Press Syndicate of the University of Cambridge
The Pitt Building, Trumpington Street, Cambridge CB2 1RP
40 West 20th Street, New York NY 10011, USA
10 Stamford Road, Oakleigh, Melbourne 3166, Australia

First published 1989

Printed in Great Britain by
Redwood Burn Limited, Trowbridge, Wiltshire

British Library cataloguing in publication data

Poole, A. (Alan)
A practical approach to toxicological investigations.
1. Medicine. Toxicology. Laboratory techniques
   I. Title   II. Leslie, G. B. (George B)
         615.9′0028

Library of Congress cataloguing in publication data

Poole, A. (Alan)
A practical approach to toxicological investigations / A. Poole,
G. B. Leslie.
         p. cm.
Includes bibliographies and index.
         ISBN 0 521 34118 3
1. Toxicity testing.   I. Leslie, G. B. (George B.) II. Title.
   [DNLM: 1. Toxicology.   QV 600 P822p]
         RA1199.P66 1989
      615.9′07–dc20   89–7078   CIP

         ISBN 0 521 34118 3

CONTENTS

FOREWORD

It is a fact of life that the disciplines of greatest current importance are the least well represented in terms of university chairs
and departments. Toxicology falls into this category. A consequence is that toxicology is a melting pot for scientists coming
from a wide range of more established disciplines including
chemistry, physics, statistics, biochemistry, biology, zoology,
physiology, pharmacology, molecular biology, pathology,
haematology, veterinary medicine, human medicine, the behavioural sciences, etc. Where groups of erudite scientists
coming from these many disciplines converge on a problem
they can, between them, if they get their act together, achieve
great heights and be seen to be on the frontiers of science.
However, before they can easily and meaningfully work
together, they have to share a common understanding of the
current state of the art and of what the general public and
regulatory authorities expect of them, and there is a special
jargon which they need to master. Unless a would-be toxicologist knows instantly what abbreviations such as QA, NOEL,
OECD, GLP and many others mean, he will be at a disadvantage when it comes to understanding what, why, and the
way in which toxicological studies are conducted. In this
context, the list of frequently used abbreviations (p. xiii) will
be much appreciated.

The aim of this book is to help scientists coming from other disciplines into the melting pot of toxicology to acquire as quickly as possible the basic information which everyone needs to have to function efficiently and effectively in the field.

The book is written by experienced senior toxicologists who are currently active in the evaluation of pharmaceutical agents from the viewpoints of both efficacy and safety. The contents of the book constitute an up-to-date crystallisation of their practical knowledge and experience.

For various non-scientific reasons, there is a tendency for studies concerned with the efficacy of a prospective drug, and studies concerned with its safety evaluation, to be regarded separately and to be the responsibility of different teams of people. A further tendency is for safety evaluation to be planned and carried out as an exercise in ticking off a check list of requirements laid down by one or other regulatory authority. Everyone working in the field of toxicology knows that this approach debases science and that research aimed at discovering the mechanisms underlying both efficacy and toxicity is much more important and much more interesting. The authors of the present book know this full well and have both contributed in an important way to the understanding of mechanisms, particularly in the field of H_2-blocking agents.

Francis J. C. Roe
D. M., D. Sc., F. R. C. Path.
Wimbledon Common

PREFACE

This book aims to provide an elementary outline of the toxicological investigations which are required in order for a drug to reach the market place. This 'child's guide' is intended primarily as an introduction to the subject, providing an insight into the worldwide regulatory environment of toxicology. As such it should be useful for undergraduates and postgraduates studying for toxicological examinations as well as graduate scientists engaged in the practice of toxicology. It should also be of value to the individuals with an interest in the conduct and assessment of toxicological studies such as clinicians, regulatory affairs, managers of small companies who use contract research facilities, teaching staff and technical advisors *inter alia*.

We would like to thank Mr David Gask for providing the figures and Mrs Janice Richardson for secretarial assistance.

Lastly A.P. is especially grateful to his wife for her forbearance during the production of the book which is dedicated to Andrew and Victoria.

A.P.
G.B.L.
May 1989

ABBREVIATIONS OF TERMS COMMONLY EMPLOYED IN TOXICOLOGY

AKP	alkaline phosphatase
ALT	alanine aminotransferase
APTT	activated partial thromboplastin time
AST	aspartate aminotransferase
BUN	blood urea nitrogen
CD	Caesarian derived
CK	creatine phosphokinase
COBS	Caesarian obtained barrier sustained
CPMP	Committee on Proprietary Medicinal Products
CSM	Committee on Safety of Medicines
CTC	clinical trials certificate (UK)
CTX	exemption from clinical trials certificate (UK)
DHSS	Department of Health and Social Security
DMPK	Drug metabolism and pharmacokinetics
DNA	deoxyribonucleic acid
ECG	electrocardiograph
ED_{50}	effective dose in 50% of animals
F1	first generation
FDA	Food and Drug Administration
GLDH	glutamate dehydrogenase
GLP	good laboratory practice
Hb	haemoglobin concentration
IND	investigational new drug application (USA)

LD_{50}	lethal dose in 50% of animals
LDH	lactic acid dehydrogenase
MCH	mean corpuscular haemoglobin
MCHC	mean corpuscular haemoglobin concentration
MCV	mean corpuscular volume
MLD	minimum lethal dose
MTD	maximum tolerated dose
NDA	new drug application (USA)
NOEL	no observed effect level
OECD	Organisation for Economic Cooperation and Development
PCV	packed cell volume
PL	product licence (UK)
PT	prothrombin time
QA	quality assurance
RBC	red blood cell count
rDNA	recombinant DNA
SCE	sister chromatid exchange
SOP	standard operating procedure
SPF	specific pathogen free
TP	total protein
UFAW	University Federation for Animal Welfare
WBC	white blood cell count

1

Introduction

The bulk of this book is devoted to the practical questions of 'what toxicological studies should we perform?' and 'how should we perform them?' In this first chapter we consider some of the more philosophical questions of toxicology; the 'whys?', the 'whens?' and the 'on whats?'

Compounds which undergo toxicity testing may be conveniently categorised as those which are intended for administration to man and those which are not. The former include pharmaceuticals to be used medicinally or prophylactically and chemicals which are added to our food, drinks or medicines to improve their stability, appearance or palatability. Those compounds which are not intended for human consumption can nevertheless represent a toxic hazard to man if they find their way into the food chain or are consumed accidentally or on purpose. This category includes biocides of various types, environmental contaminants, industrial chemicals, veterinary products, animal feedstuffs, food additives, household chemicals and natural toxins.

Since it is on pharmaceuticals that the most comprehensive toxicological evaluations are generally performed, this book has been directed primarily towards the toxicological evaluation of potential new drugs. The principles and methodology of toxicological evaluation of other types of compounds are essentially similar. We shall not, in this book, consider aquatic and environmental toxicology.

1.1 THE PURPOSE OF TOXICOLOGICAL TESTING

It was recognised over 400 years ago, by a remarkable character, Phillipus Theophrastus Bombastus von Hohenheim (1493–1541) who called himself Paracelsus, that all compounds are toxic. In his Third Defence he said: 'All substances are poisons: there is none which is not a poison. The right dose differentiates a poison and a remedy.'

The role of the toxicologist is to investigate and establish the nature of such toxicity in order to help assess human risk. The purpose, therefore, is not to demonstrate 'safety' or show that a compound is non-toxic. Indeed a toxicologist should suspect that he may have failed in his task if toxicity is not demonstrated! Thus toxicologists are not primarily concerned with whether a chemical is toxic but with how toxic a compound is and, with potential drugs, the targets for its toxicity and ratio between the toxic dose and the therapeutic dose. This 'therapeutic ratio' can be expressed in a variety of ways. One can divide the LD_{50}* by the ED_{50}† or, perhaps better, divide the LD_1 by the ED_{99}. More meaningfully, the maximum dose found to be free from the adverse effects in long term toxicity studies should be divided by the highest estimated therapeutic dose. Such ratios help us to make risk–benefit comparisons between potential drugs. The acceptable ratio will, in practice, vary with the clinical indication. A drug intended for the treatment of a life threatening condition would be acceptable with a much lower ratio than one intended for the alleviation

---

* LD_{50} (lethal dose in 50% of animals) is the dose which will cause mortality in 50% of the animals in a test population.

† ED_{50} (effective dose in 50% of animals) is the dose which produces desired response in 50% of animals.

of a minor complaint. With compounds which are not potential drugs risk assessment will be based on rather different criteria, but the principle of risk–benefit considerations will nevertheless still apply.

1.2 PREDICTION

As well as investigating the toxicological nature of a compound in animal and *in vitro* studies, the toxicologist must attempt to determine whether the adverse effects seen in such experimental systems are predictive for man. If they are, then it becomes important to know at what dose and length of exposure they may become apparent, and also whether there are any factors which may predispose an individual to show such effects. This implies that the toxicologist must go much further than merely identifying the organ or systems at risk, and must also study the mechanism of the adverse effect. Unfortunately, all too often, toxicologists confine themselves to describing the natural history of a toxic event which, although important, is a poor substitute for understanding the mechanism of toxicity. This need for a more scientific approach to toxicology was highlighted in 1978 when a report by the Royal Society (*Long-term Toxic Effects – A Study Group Report*) stated that 'toxicology is not adequately supported as an academic discipline; it is swamped by routine tests of limited value and governed by regulations rather than by rational thought. This activity tends, however, to be diverted into a sterile routing of exposure of many animals, without attention to mechanisms. Only if mechanisms are understood is it possible to extrapolate toxicity measurements across species ...'

Animal studies using high doses of compounds over long periods of exposure will inevitably produce adverse effects which are 'false positives' in the sense of not being predictive of a similar adverse effect in humans. Unless we are willing to 'throw out the baby with the bathwater', we cannot readily accept that an adverse effect seen in experimental animals (or *in vitro*) means that we should abandon work on an otherwise promising potential drug. Conversely sometimes the adverse effects may be such that a compound can no longer be considered for use in man, e.g. evidence of genotoxic carcinogenicity. Much more often, however, it is necessary to undertake investigative studies to analyse the mechanism of the adverse effect or evaluate its species specificity. It is quite often possible to demonstrate that a toxic effect is specific to a particular animal species, thereby being non-predictive of an adverse effect in humans. *In vitro* techniques have often been of particular value in this respect since it is possible to compare the toxic effects of drugs on cells (primary or continuous cultures) derived from human or animal tissues. There is no doubt that in the future the use of cell culture techniques in toxicology will become much more common. Two new journals devoted to this field have already commenced publication, a number of international symposia have been held and several books have reviewed the field, e.g. Atterwill & Steele (1987).

The mechanistic approach to toxicology may be expensive in terms of time and scientific effort, but is obviously more logical and cheaper than abandoning work on a potentially valuable new medicine. At the very least, investigations into the mechanisms of toxic effects may help to establish short term screening procedures useful in selecting 'backup' compounds lacking the undesired effect. Scientific or mechanistic toxicology therefore has a valuable role to play in the drug design process. In order to carry out investigative studies it is of course necessary to employ a number of specialist scientists from a range of disciplines. As it is impracticable, even in a large organisation, to cover all specialities it is necessary to have recourse to academic specialists and therefore close ties between industry and academia are essential.

A wide variety of scientists consider themselves to be toxicologists and there has been much debate as to who is best equipped to conduct toxicological studies and how toxicologists should be trained. In dictionaries toxicology is often defined as the 'science of poisons'. This definition is, however, quite unsatisfactory since toxicology is not a single science. In order to examine the toxic potential of a novel agent thoroughly, it is necessary to have a team of scientists from a range of disciplines, with initial training in human or veterinary medicine, biochemistry, physiology, pharmacology, pathology, genetics or other biological sciences. Training in toxicology is obviously important, but probably best carried out as postgraduate study with a considerable element of practical training.

1.3 STUDY DESIGN

In an attempt to be predictive we must select, if possible, animal or *in vitro* models which have been shown historically to have value and relevance to humans. Although it is not possible to say that any particular species is most like man, an important consideration in toxicity studies is to try to choose the species which metabolise the compound in a similar manner and with similar pharmacodynamics to man.

This desire to use a species with similar drug metabolism to that of man produces something of a chicken and egg paradox. In order to study drug metabolism and pharmacokinetics in humans it is first necessary to have completed toxicity studies in experimental animals which may have a completely different metabolic profile from that of man. Naturally, once the human studies have been carried out it may then become possible to be more rational in the choice of species for longer term studies. Another approach, before beginning any animal studies (or, naturally, human investigations), is to study drug metabolism in primary cultures of hepatocytes isolated from animal and human liver. Such a comparison of *in vitro* metabolism may assist in selecting the most appropriate species for toxicity studies. It must be remembered, however, that despite the desire for comparability between metabolism in the experimental animals and man, choice of species of use in toxicology studies is limited. It may therefore be necessary, in some instances, to conduct animal toxicity studies on a human metabolite if that metabolite is not produced in the experimental laboratory species.

Another consideration is that the numbers of animals used in toxicology studies are necessarily very small compared with the human population which may be exposed to the drug. Furthermore, most experimental animals are genetically very homogeneous, often living in closely controlled environments with uniform diets. So far as possible all other biological variables such as age, weight, health status, diurnal and seasonal variations are minimised. This is in sharp contrast to the situation in humans who are genetically very diverse and live in a wide variety of cultures and environments with enormous ranges in diet, nutrition, health status, age and weight. Yet we hope to make predictions from data obtained from small populations of homogeneous laboratory animals applicable to the enormous diversity of the human population in general.

This expectation is based at least in part on a devout belief in the relationship between dose and response. Although there are types of adverse drug effects which appear to be unrelated to dose, most effects become more prevalent and more pronounced at higher dose levels. Thus it is hoped that a frequent event seen at a high dose given over a long period of time to homogeneous experimental animals will be predictive of a rare event produced at much lower doses in a small number of individuals in the vast and diverse human population. Documentary evidence to support this faith in the predictive value of toxicological studies is not readily available, but a recent attempt to consider how predictive are toxicity studies was undertaken by Lawrence, McLean & Weatherall (1984). Six compounds had their toxicological findings reveiwed in the light of clinical experience with the drug. Further 'case histories' comparing animal toxicology study predictions with clinical experience are essential if we are to validate and improve our toxicological assessments.

A major reason for constructing dose–response curves is to establish a maximum no-adverse-effect level, which generally provides some feeling of comfort for the researcher undertaking clinical investigations and satisfies regulatory authorities. However, regulators, in general, have not accepted the argument for a no-effect level for compounds which prove to be oncogenic at high doses, and normally consider that there is no safe dose for a carcinogen. Such a position is, however, clearly scientifically untenable since it is now clear that chronic irritation can cause tumours as can interference with endocrine feedback mechanisms. Clearly doses which do not cause these physical and/or physiological changes will not be oncogenic. Recently Ames (1987) stated that over 50% of chemicals tested in oncogenicity studies in animals could be classified as carcinogens, but that most of these do not appear to act by damaging DNA. He also pointed out the ubiquity of carcinogens which we readily accept in our diet, e.g. parsley, basil, peanut butter, beer, cola, bread and wine. Based on such observations, a positive finding of oncogenicity at a high dose level should not be taken to be necessarily predictive of an oncogenic potential in man at clinical dose levels.

1.4 SOCIAL AND LEGAL CONSIDERATIONS

Besides the evident scientific, ethical and commercial motivation to carry out toxicological investigations, there are nowadays regulatory requirements to be fulfilled. (These are discussed in detail in Chapter 2.) The changing regulatory climate and the introduction of 'good laboratory practice' have had a major impact on the modern approach to toxicological investigations. There has been a need for additional skilled manpower and an increased awareness of the need for technical and scientific training, especially in the areas of computerisation and 'on-line' data collection and processing.

Not least of the changes has been the formalisation of the standard packages of studies which are nowadays expected. In this rather rigid climate, it is all the more necessary to motivate and maintain the interest of high calibre scientists so that a flexible investigative approach to any adverse effects found in the 'regulatory' studies may be achieved.

Other factors which influence toxicology are the increasingly vociferous pressure from anti-vivisectionists and recent changes in the law in the UK and other countries relating to animal experiments. The majority of standard 'regulatory' toxicity studies are still carried out on conscious animals and will continue to be done so for the foreseeable future. No *in vitro* systems are able to provide adequate substitutes for the complexities of the intact body with its interactive mechanisms and feedbacks. *In vitro* systems cannot predict in detail the absorption, metabolism, distribution and elimination of compounds.

However, many toxicologists are striving to minimise the numbers of animals used in studies by maximising the data obtained from each animal and by improving experimental design. Where replacement of animals by *in vitro* systems does offer viable and validated alternatives, they have been rapidly adopted. Mechanistic toxicology and the screening of candidate compounds to avoid a previously identified type of adverse effect are areas where *in vitro* methods have been widely used.

Toxicology is a continually evolving science still undergoing rapid and far reaching changes. Toxicologists approaching this science with a thorough understanding of their role and a high level of technical competence should be able to adapt to, and indeed be instrumental in, producing these changes. The purpose of this book is to provide the fundamental knowledge, in a succinct and easily understood way, which will allow the beginner to understand the basic theory and practice of toxicological investigations.

REFERENCES

Ames, B. N. (1987) Six common errors relating to environmental pollution. *Regulatory Toxicology and Pharmacology*, 7, 379–83.

Atterwill, C. K. & Steele, C. (eds.) (1987) *In Vitro Methods in Toxicology*. Cambridge University Press.

Lawrence, D. R., McLean, A. G. M. & Weatherall, D. M. (1984) *Safety Testing of New Drugs. Laboratory Predictions and Clinical Performance*. Academic Press, London.

The Royal Society (1978) *Long-term Toxic Effects – A Study Group Report*. The Royal Society, London.

2

Regulatory requirements

The self-evident goal of pharmaceutical companies is to develop and market high quality, efficacious and safe drugs. The consequences of marketing drugs which have been inadequately evaluated for safety could be so damaging to a company that self regulation should be adequate to ensure that the highest standards of testing are achieved. However, while most companies have worked to very high standards (both scientifically and ethically) there have been, unfortunately, occasions when such ideals have not been achieved. As a result of such lapses legislation has been introduced in an endeavour to ensure that safety testing should reach acceptable minimum standards.

2.1 PRECLINICAL SAFETY TESTING LEGISLATION

The Federal Food, Drug and Cosmetic Act introduced in the USA in 1938 is the oldest law introduced to regulate the exposure of humans and animals to drugs, medical devices and cosmetics. The original act has been amended several times and now requires that drugs must be both effective for the labelled indication and safe. In the USA, the Food and Drug Administration (FDA) is responsible for the safety testing of therapeutic agents and, in common with other regulatory authorities, has specified programmes for preclinical toxicity testing.

In the UK the first regulatory agency, established in response to the thalidomide tragedy, was a voluntary committee (Committee on the Safety of Drugs) set up by the pharmaceutical industry and chaired by Sir Derrick Dunlop. However, the Medicines Act of 1968 gave statutory powers to a Committee on the Safety of Medicines. Many other countries have now also passed laws regulating drug safety testing. While some authorities have opted for systems of recognised 'experts' to examine submitted safety data, others, such as the USA, prefer formal committees.

In order to support an application for the registration of a new drug it is necessary to satisfy legislation requiring that certain data be produced from a variety of toxicological investigations. However, in the majority of cases the legislation does not precisely describe what preclinical toxicity studies should be conducted. For instance, to market a drug in the USA it is necessary 'by all means reasonably applicable' to demonstrate to the FDA that a compound is 'safe', the German Medicines Act (1976) requires that documentation for registration has 'results of pharmacological and toxicological testing', while the UK Medicines Act (1968) requires that applications should be accompanied by 'information and documents regarding the safety of medicinal products'. The reason for this 'lack of specificity' is based upon the assumption that safety studies should be based upon 'state-of-the-art science' which can evolve more rapidly than laws can be changed. Therefore rather than establishing laws the majority of authorities have preferred to announce their requirements for particular types of toxicity studies in the form of generic standards or guidelines. (Some of the major regulatory authorities together with some of the major national and international guidelines for toxicity studies are shown in Table 2.1.)

Table 2.1. *National regulatory authorities and guidelines*

Country	Regulatory authority	Guidelines
Australia	Australian Department of Health	Guidelines for Preparation and Presentation of Applications for Investigational Drugs and Drug Products (1987)
Canada	Health Protection Board (HPB)	Preclinical Toxicology Guidelines (1981)
France	Ministry of Public Health and Social Security (MPHSS)	Guidelines for the Analytical, Pharmacological and Toxicological Testing of Pharmaceuticals (1976)
Japan	Ministry of Health and Welfare	Guidelines for Toxicity Studies of Drugs (draft) (1988)
Nordic countries	–	Guidelines for Registration of New Drugs (1983)
UK	Department of Health and Social Security (DHSS)	Guidance Notes on Applications for Product Licences (MAL2) 1986 (Supplement 1987)
USA	Food and Drug Administration (FDA)	Guidelines issued by the PMA but prepared in conjunction with the FDA (1977)

A major point stressed by the regulatory authorities is that the guidelines are flexible, and not specific requirements, thus giving individual scientists the freedom to consider, and if necessary use, alternative approaches. Nevertheless despite these 'flexible guidelines' many authorities are routinely criticised for not being logical enough in their approach; being more concerned with fulfilling preconceived requirements, no matter how unreasonable or unsuitable for the drug in question, rather than accepting detailed, measured, scientific arguments. Thus it is not easy to know whether a study which deviates from the 'standard acceptable procedures', even for sound scientific reasoning, will be acceptable to certain authorities.

A number of international organisations and committees such as the Organisation for Economic Cooperation and Development (OECD) and the Committee on Proprietary Medicinal Products (CPMP) have been established to help harmonise the data requirements and procedures for different countries, with regard to regulation of drugs. Nevertheless there are national idosyncrasies which seem in some cases to operate as *de facto* non-tariff trade barriers. Minor variations in requirements, while not prejudicing patient safety, may effectively retard or even prevent marketing of a drug in a particular country. Thus it behoves toxicologists involved in developing drugs for worldwide marketing to be conversant with the full international range of regulatory requirements in order to ensure that their protocols cover, where possible, all the various requirements and avoid the costly exercise of having to repeat studies.

For convenience we have categorised the regulatory requirements into general toxicology, reproductive toxicology and genotoxicity studies (summaries are shown in Tables 2.2, 2.3 and 2.4 respectively). These tables, while not attempting to provide a comprehensive survey of national requirements, should serve to give some indication of the types of study which regulatory authorities wish to see before giving approval for clinical trials and before registration of new drugs. The reader should, however, bear in mind the fact that guidelines are constantly being updated and what is considered suitable today may not be so acceptable tomorrow. (Fuller descriptions of the methods used to fulfil regulatory requirements are described in subsequent chapters and in Appendix 1 and, where there are major differences in the requirements of different authorities, these are also discussed.)

Table 2.2. *Summary of regulatory toxicology requirements*

	Single dose (acute study)	Repeat daily dosing 2 weeks – 3 months (sub-chronic study)	Repeat daily dosing 6–12 months (chronic study)	Oncogenicity study repeated daily dosing (minimum) rats: 24 months; mice: 18 months
Animals				
Species	At least two species usually rodents but certain authorities may demand third non-rodent species	Two: (i) rodent, usually rat (ii) non-rodent, dog or monkey	Two: (i) rodent, usually rat (ii) non-rodent, dog or monkey	Two rodents, usually rat and mouse
Group size	Varies since high precision of accuracy for estimating LD_{50} not usually required	5 to 10 per sex per group rodent At least 3 per sex per group non-rodent (3 dose groups, 1 control group)	At least 20 per sex per group (rodent) At least 5 per sex per group non-rodent (3 dose groups, 1 control group)	At least 50 per sex per group (now common to have a 5 group study design, i.e. Control (group 1) Control (group 2) low, mid, high doses group 3, 4 and 5 respectively
Treatment	Two routes of administration, (one via proposed clinical route and second usually intravenous)	Proposed clinical route	Proposed clinical route (if possible)	Gavage dosing, unless contra-indicated by other factors

In-life studies				
Clinical observations	Daily	Daily	Daily	Daily
Body weight	–	Before dosing then weekly	Before dosing then weekly	Before dosing and then weekly
Food and water consumption	–	Before dosing and at times during study	Before dosing and at times during study	Selected times during study
Ophthalmoscopy	–	Beginning and end of study	Beginning and end of study	Beginning and end of study
Clinical chemistry	–	Usually beginning, middle and end of study	Usually 4 to 5 times during course of study	–
Haematology	–	Usually beginning, middle and end of study	Usually 4 to 5 times during course of study	–
Urinalysis	–	Usually beginning, middle and end of study	Usually 4 to 5 times during course of study	–
Terminal studies				
Autopsy	Full autopsy, abnormal tissues taken	Full autopsy, all tissues taken. Major organs weighed	Full autopsy, all tissues taken. Major organs weighed	Full autopsy, all tissues taken
Histopathology	On abnormal tissues	On all tissues	On all tissues	On all tissues

Table 2.3 *Summary of regulatory requirements for reproduction studies*

	Teratology study	Fertility and general reproductive performance study	Peri- and postnatal study
Animals			
Species	At least 2 mammalian species, 1 rodent (usually rats) and a non-rodent (rabbit)	Rat or mice	Rat or mice (usually rat)
Group size	At least 20 pregnant female rodents At least 12 pregnant female non-rodents (3 dose groups, 1 control group)	At least 24 per sex per group (3 dose groups, 1 control group)	At least 20 pregnant animals (3 dose groups, 1 control group)
Treatment	Throughout embryogenesis rodent [a]P6–15 non-rodent (rabbit) P6–18 (Proposed clinical route of administration)	Males: 70 days prior to mating Females: 14 days prior to mating until autopsy (Proposed clinical route of administration)	From P15 and throughout lactation to day 21 (weaning) (Proposed clinical route of administration)
In-life studies			
Clinical observations	Daily	Daily	Daily
Body weight	Daily	—	—
Food consumption	Daily	—	—

Terminal studies			
Time of necropsy	Dams killed before parturition	Males: after mating Females: $\frac{1}{2}$ on P22, remainder after weaning	Dams and litters at weaning, i.e. day 21, or when offspring reach sexual maturity. (Some countries require information on the reproductive capacity of the offspring and may even demand information on the F_2 generation – for further information see Chapter 8)
Foetal examination	Rodents: $\frac{1}{3}$ to $\frac{1}{2}$ foetuses visceral examination $\frac{2}{3}$ to $\frac{1}{2}$ foetuses skeletal examination Rabbit: All foetuses visceral and skeletal examination	As for teratology study	
Examination of litters			
Body weight Post-natal development and reproductive function	Usually no F_1 offspring produced. However for certain countries e.g. Australia and Japan a percentage of dams should be allowed to litter normally and the offspring subsequently examined. (For further information see Chapter 8)	Weekly Assessment required but no firm details given (for further information see Chapter 8)	Weekly after weaning Assessment of offspring for visual, auditory, behavioural and reproductive functions

(a)P = day of gestation.

2.2 GOOD LABORATORY PRACTICE (GLP)

Apart from the necessity of performing appropriate safety tests on new compounds, it is of course essential that such studies are conducted to a rigorous scientific standard and that hazard assessments are based upon data of high quality and reliability. The need for GLP was highlighted by the FDA who reported that a number of preclinical laboratory studies which had been performed to investigate the toxicity of novel compounds were deficient in a variety of ways, e.g.:

 (i) studies poorly conceived and executed with certain analyses inaccurately performed and/or reported;

 (ii) personnel not adequately trained and also, in some instances, not adhering to test protocols;

(iv) critical review of data and proper supervision of studies not always carried out.

While most laboratories were found to conduct studies rigorously and conscientiously, it was decided that the problem was sufficiently serious to warrant the introduction of legislation which should prevent future deficiencies. The

Table 2.4 *Summary of regulatory requirements for genotoxicity studies*

---

1. Test for gene mutations in bacteria (Ames test).
2. Test for chromosomal aberrations in mammalian cells *in vitro*.
3. Test for gene mutation in mammalian cells *in vitro*.
4. *In vivo* test for genetic damage.

The criteria for selection of these assays are in agreement with the description of four categories of tests cited in the Notes for Guidance for the Testing of Medicinal Products for their Mutagenic Potential (EEC, Notice No. 84/C293/04 Annex II). These tests, perhaps with minor protocol modifications, e.g. inclusion of *Escherichia coli* (*E.coli* wP2 uvrA trp) in the bacterial mutation test to satisfy Japanese Regulations, are now widely used and accepted for worldwide clinical testing and registration of drugs.

For a full description of these tests, see Chapter 9.

purpose of GLP was therefore to ensure the quality and integrity of the test data submitted as a basis for regulatory decisions on drugs. The rationale was that well-founded GLP standards would contribute to improved protection of health.

The first proposals for GLP were made by the FDA in 1976 and were finally published as law in 1979. Although the FDA, as a US authority, had no legal authority in other countries, any company wishing to market its products in the USA had to comply with the FDA GLP requirements. As a result of this, FDA inspectors visited a number of laboratories in Europe before European countries began to follow the US lead and introduced their own GLP compliance programmes and inspections. Thus in a very short period of time GLP has changed from being a requirement for a single country to being necessary internationally. Although many countries now have their own GLP requirements, there are attempts to harmonise standards and requirements so that a national certification of compliance will become internationally acceptable.

As might be expected, GLP regulations although not designed to be 'over-detailed and constrictive as to prevent the full deployment of scientific initiative, experience, expertise and judgement', do provide for very detailed instruction about the organisational process of non-clinical laboratory studies and the conditions under which they are planned, performed, recorded and reported. GLP is not therefore concerned with which studies should be performed, but only with how such studies are carried out, monitored and recorded. The scope of GLP is very wide and has had an immense impact on the conduct of toxicology studies. It is therefore essential to appreciate, in outline at least, some of the principles and procedures with which laboratories conducting preclinical studies must comply. Examples of such principles include:

> *Test facility and personnel*: organisations must ensure that they have facilities of suitable size, construction, location, etc. to meet the requirements of study. All

personnel involved in studies must have adequate training and experience (training records and *curricula vitae* of all staff must be kept).

Apparatus: sufficient apparatus of adequate quality and capacity must be available to meet the requirements of the test. Also, apparatus must be working correctly and not producing spurious results etc. (documentation of calibrations and servicing should be kept).

Test substances: provision should be made for procedures designed to ensure the identity, purity, stability, composition etc. of test materials to be used in studies.

Standard operating procedures: all laboratory facilities must have written instructions for any procedures which may be used in studies (such instructions should cover all procedures, from the simplest to the most intricate of scientific techniques).

Protocols: all studies must have written protocols and any deviation from that protocol must be recorded in appropriate numbered amendments.

Records and reports: archives for storage of study plans, raw data, specimens and final reports must be provided. Reports must include certain information such as objective of study, full details of the composition and stability of the test substance, a description of the test system in use, summary of findings, etc., and must be signed by the study director and other senior staff and consultants.

Quality assurance programme: companies must provide a mechanism by which the management is assured that studies are being conducted in compliance with GLP procedures. A quality assurance manager is usually required to demonstrate that the system is working.

Authorities which require compliance with GLP usually provide a GLP Monitoring Unit which not only gives information, advice and guidance to companies but will also inspect laboratories to monitor the testing of compounds. If the inspection unit find that procedures used in laboratories are not acceptable then data generated in such facilities may prove unacceptable to regulatory authorities.

It must be emphasised that the above list is only a very brief summary of what is required to comply with GLP. Anyone requiring more detailed information should obtain general recommendations for GLP from the appropriate governmental body, e.g. United Kingdom Compliance Programme issued by the Department of Health and Social Security in 1986; or non-governmental bodies, e.g. OECD Principles of Good Laboratory Practice, Implementation of Good Laboratory Practice and Guidelines of National GLP Inspection and Study Audits.

There is no doubt that GLP has been instrumental in preventing many of the malpractices operating in some laboratories prior to its introduction. Thus it has achieved its ultimate purpose of protecting health. Unfortunately there has been a tendency to make GLP complex and cumbersome, while the multiplication of GLP regulations has made international harmonisation difficult. It is to be hoped that eventually there will be soundly based, practicable GLP standards which are not unnecessary and do not put excessive burdens on those responsible for conducting toxicological investigations.

2.3 REQUIREMENTS FOR CLINICAL TRIALS AND MARKETING

For drug development it is essential to know, as early as possible, if the test agent will produce the desired pharmacological effect in man and also, as discussed previously, to have information on human kinetics and metabolism. It is therefore necessary that toxicology and clinical studies parallel one another, with of course the clinical investigations running behind the toxicological investigations. Data from toxicity studies also give the necessary information on whether to proceed to clinical trials, since they provide knowledge of potential target organs and tissues and some understanding of the mechanism of action.

Clinical investigations can be divided into phases with each phase being dependent upon the quantity and quality of toxicological data already available. Brief definitions of the different phases are listed below:

Phase I: Volunteer studies in healthy subjects to examine absorption metabolism, excretion, pharmacokinetics, pharmacological responses etc.

Phase IIa: Studies in patients requiring treatment, to examine the clinical pharmacological aspects of the material. Studies can proceed from single to multiple doses.

Phase IIb: Studies in patients to examine the efficacy of the drug. These studies are usually quite large involving repeated dosing.

Phase III: Long term safety and efficacy studies prior to marketing.

Phase IV: Postmarketing.

Unfortunately the majority of countries provide only minimal guidance on the data required for clinical trial approval. Whereas the legal position varies from country to country, in general the length and type of clinical trial allowed depends mainly upon the type of toxicology data available. (Some examples of the length of repeated dose toxicity correlated to duration of human exposure for different authorities is shown in Table 2.5.)

There are also differences between regulatory authorities in not only the quantity of data they require, but in how (or indeed whether) such data are examined before clinical trials are allowed. In the UK it is possible to apply for a clinical trials exemption certificate (CTX) or a clinical trials certificate

Table 2.5. *Examples of requirements of toxicity data in respect of clinical trials*

Duration of human administration	Duration of toxicity study in experimental animal (2 species) required by different regulatory authorities				
	Sweden	CPMP	UK	Canada	USA
Single dose or several doses on 1 day	2–4 weeks	2 weeks	2 weeks		2 weeks
Repeat dosing up to 7 days		4 weeks	4 weeks	4 to 6 months	
Repeat dosing up to 30 days	At least 3 months	3 months	3 months	18 months	1 month
Repeat dosing beyond 30 days	At least 6 months	6 months	6 months		3 months

Clinical trials may be carried out in men without any reproductive data. However, to include women it is often necessary to have completed teratology and female fertility studies.

(It must be remembered that this table provides only a summary guidance and that requirements may be changed depending on the properties of the drug under investigation.)

(CTC). The former is a negative vetting system in which, in theory at least, if the responsible physican has signed the clinical trials application there is no need to review all of the preclinical data. Reliance is therefore placed upon the professional integrity of the company, the recommendations of the medical review committees operating within the company and the clinical judgement of the physicians in the public sector who will be conducting the trials. In practice, however, summaries of preclinical biological studies, e.g. toxicology, pharmacology, pharmakokinetic, etc., are also supplied and can be used to make decisions concerning clinical development. The idea of the CTX was to speed up the development process by allowing clinical trials to proceed within about 30 days of applying for the certificate. With a CTC all preclinical data, not just summaries, are supplied and all the documentation is reviewed before trials may proceed. Since this procedure can take up to six months, it is not the first choice of companies wishing to develop new medicines. Of course, if queries arise concerning the summaries supplied in support of a CTX application, full reports may be requested.

In order to conduct clinical trials in the USA it is necessary to apply for an IND (investigational new drug application). To grant this the FDA requires all data (as in a CTC) to be submitted and reviewed before trials can begin. With authorities in some other countries, while it may be necessary to file preclinical data (reviews and/or full reports) before beginning clinical trials, these are not usually examined. The files are only opened in the event of adverse effects occurring in the patients taking part in the trials.

A particular concern at the early clinical trial stage may be possible reproductive toxicological effects. These problems can be largely avoided simply by precluding women from the studies, or women of childbearing age. (Clinical trials can be carried out in men without any reproductive data.) Eventually, if the drug is to be prescribed to women of childbearing age, it is necessary to provide reproductive toxicity data. It is, however, possible in certain countries to undertake Phase IIa clinical trials in women of childbearing age having completed teratology studies (two species), e.g. in the UK and Australia. The full package of reproductive studies (see Chapter 8) are, however, required for the therapeutic trials (IIb and III). Also, before administration to man, some countries, such as the UK, demand that chemicals have been examined in a genotoxicity assay (see Chapter 8). Intravenous/perivenous irritancy studies and perineural studies may also be required prior to clinical trials with intravenous formulations.

There are several reasons for carrying out clinical trials as early as possible. These include determination of possible differences in drug metabolism between man and the experimental animal, the effectiveness of the drug in man and an idea of the anticipated human dose. This latter point is most important to the toxicologist since chronic studies rely on giving dose levels which are multiples of the pharmaco-logically effective dose and proposed human dose.

Clinical trials should provide:

Pharmacology data: dose–response relationships, thera-peutic index, optimum dose, etc.

Pharmacokinetic data: absorption, bioavailability, distribution, metabolism, protein binding, excretion, etc.

Drug interactions: important for drugs with high protein binding, metabolised extensively in the liver (effect of metabolic inducers). Effect of concomitantly pre-scribed drugs on blood levels of test material.

Therapeutic efficacy: well designed, double blind trials, controlled with a placebo and/or another approved drug. Demonstration of a physiological effect is not sufficient, drugs must show symptomatic relief.

Safety data: haematology, clinical chemistry, liver and kidney function tests. Patients closely observed for drug interaction, dependence, etc. Clinical trials may have to include special categories of individuals, e.g. the elderly, pregnant women, etc. Passage to the breast milk may be investigated if the drug is intended for nursing mothers.

In all studies the test drug should be standardised with respect to identity, strength, quality and purity. Also the for-mulation used in clinical trials should be the same as that intended for marketing.

It is difficult to be precise about the number of patients, in total, which will be needed to support a marketing appli-cation. In the UK, data on 500–1000 patients may be required for some types of drug. For all countries the number of patients on which data will be required depends upon the nature of the condition the drug is directed against, alternative treatments available, findings in preclinical toxicology studies and issues arising from the clinical studies.

2.4 PRECLINICAL SAFETY TESTING OF PRODUCTS FROM BIOTECHNOLOGY (rDNA)

A problem facing pharmaceutical companies and regulatory authorities alike is what preclinical toxicity studies should be performed on those products produced by recombinant DNA (rDNA) techniques or hybridoma-derived monoclonal antibodies. Because of lack of experience in the safety testing of rDNA products there are, at the time of this book going to press, no validated or widely accepted methods of safety testing. A number of authorities have, however, issued draft recommendations and while they are mainly concerned with production and quality assurance they do show some common attitudes towards toxicity testing in laboratory animals. A summary of some common points is listed below:

> *Relationship of rDNA product to natural agent*: toxicology studies are expected on those rDNA products which differ in any way from their natural counterpart. Less comprehensive toxicity testing, possibly more related to pharmacological activity, would be expected for these rDNA products identical to the natural material.

> *Species to be used in toxicity studies*: toxicity in *one relevant* species. This species should be one in which the pharmacological activity of the rDNA compound can be demonstrated to be the same as in man, or other good biological reasons should be given for its selection. All guidelines appear to place great emphasis on testing for neutralising antibodies since it is feared that the absence of a toxic effect may simply be due to the compound being 'neutralised' by antibodies.

> *Dose levels*: it is suggested that only two dose levels be used; one at the therapeutic level with the other ten times higher. (Dose selection may be complicated by

the fact that many rDNA products will be used in replacement therapy, thus a 'therapeutic dose' in an animal not suffering any deficiency could be considered as an 'overload' dose.) It is recommended that the naturally occurring product should be included if possible as a control.

Length of studies and parameters to be measured: recommendations for duration of exposure are not well formulated and contain general statements such as 'duration and route of administration should be similar to man'. Naturally in any study the length of exposure in the experimental animal may very well depend upon the antigenicity of the compound. (Some guidelines recommend carcinogenicity studies, but it is unrealistic to believe that it will be possible to dose rDNA products to animals for two years without producing some type of immunological response.) Most recommendations suggest that, while standard parameters are also measured, increased emphasis should be placed on monitoring in-life physiological observations, e.g. blood pressure, electrocardiography and/or other cardio-vascular measurements. Attention should also be paid to determination of immunotoxicity. In most cases emphasis is placed on safety pharmacology rather than toxicity.

Additional testing: if there is clear evidence of maternal toxicity some type of reproductive toxicity programme should be undertaken. Naturally modified protocols with shorter treatment periods may be necessary if the rDNA product is antigenic.

Mutagenicity studies are also likely to be required, especially on those materials which differ significantly from the natural product. It has been suggested that mutagenicity tests may provide a sensitive method for detecting impurities in the recombinant products, e.g. residual cellular DNA or viruses. It is, however, more likely that modern analytical techniques will prove more useful for such measurements.

2.5　CONCLUSION

This review of the various regulations for the preclinical evaluation of new drugs shows that such regulations and guidelines have become complex and burdensome and that bureaucracy is in danger of stifling the pharmaceutical industry and wasting intellectual resources. With the obvious risks to the public if safety evaluation is inadequate, it is clear that such regulatory mechanisms are here to stay. Ever increasing demands for more and more regulatory requirements, which use up resources, manpower and money, and result in the use of greater numbers of experimental animals, will be seen to do little for the improvement of risk assessment. Indeed they could have a negative effect on the development of much needed improvements in many areas of drug therapy and prophylaxis.

Common criticisms of regulatory authorities are not so much with the demand for information, but the delays in processing applications and issuing product licences and the obsession with 'purely bureaucratic devices'. In many ways it would appear that authorities, rather than assisting companies to develop much needed drugs quickly and safely, often adopt an adversarial role.

It is hoped that the great efforts expended by companies to explain observed toxicological effects in mechanistic terms will be respected and understood by the authorities. Also it is hoped that a more 'scientific approach' will be adopted and observations in animals can be rationally and scientifically explained in terms of species-specific effects and not simply misused to make arbitrary, unjustified comparisons between effects seen in animals and man. Acceptance of rational, well reasoned scientific approaches will allow responsible companies to thrive and more quickly develop the drugs which are so evidently needed. It is after all a requirement of the scientist to improvise not compromise.

3

Animals: sources, selection, husbandry

Before toxicological studies can begin, experimental animals must be obtained (usually from commercial suppliers), acclimatised, checked for general health (possibly including blood tests), allocated to groups, numbered, labelled, caged and fed. The overall objective is to select and maintain animals in the best possible conditions of health throughout the course of a study. This is done for both humane reasons and also to provide the maximum opportunity for detecting and investigating adverse effects. It is also essential to minimise variables in the animals' environment which can influence the development of pathological changes. The importance of environmental conditions was demonstrated by Riley (1975) who reported that mice in a noisy environment developed mammary tumours much earlier than their siblings kept in quiet surroundings. The methods for achieving these objectives, i.e. maintaining healthy animals in a well controlled and standardised environment, can vary depending on the species being used.

3.1 RATS AND MICE

Rats and mice of various well-known strains, bred to very high standards, are now readily obtainable from a number of large commercial breeders some of which are multinational. Thus while a few organisations breed their own animals, the majority purchase them from commercial suppliers as and when required. The most widely used strains of rat in the UK are Sprague–Dawley and Wistar, the latter is a slightly smaller animal which may offer an advantage for some studies. Mice used in toxicity studies are usually commercially available inbred strains such as the CDI or C57BL but some researchers prefer to use hybrids for oncogenicity studies, e.g. B6C3F1.

Laboratory rodents, like other animals including man, can suffer from a variety of diseases. Clearly the presence of disease, which may result in the early death of a proportion of animals, can severely compromise a toxicity study. Even if the animals are examined prior to the start of the investigation, there is always the possibility that they may have a subclinical or latent disease which could go unnoticed. It is possible that if animals receiving treatment are more stressed than their counterparts in the control group this stress could activate the latent disease, giving a false impression of the toxicological properties of the test agent. For these and other reasons, over the past 20 to 30 years there has been a tendency to use experimental animals of a known or stated health status. There are now a number of terms, most unfortunately being imprecise, attemping to describe the quality of the animals. At the lowest end of the health quality scale there are the conventional laboratory animals, which must be free from diseases transmissible to man, with various grades leading to the specific pathogen free (SPF) animals which are not only free from common diseases, but also free from all pathogenic organisms known to cause disease in the species concerned. Such SPF animals are usually Caesarian derived and then maintained under special barrier conditions (see below) to ensure that they are not exposed to unwanted pathogens.

Between the SPF and conventional animals there are various degrees of health status, e.g. barrier maintained animals which, although not SPF, are kept in well-controlled environments. Other categories include gnotobiotic animals which, although not necessarily free from pathogens, carry only known organisms.

To minimise biological variation it is usual, when conducting toxicological studies, to use genetically homogeneous SPF animals. As stated earlier, such animals can be purchased from commercial breeders, and in order to ensure that they remain free from infection most modern laboratories maintain SPF rodents under barrier conditions. Inside the barrier animals breath filtered air, drink filtered water and eat sterilised food. All staff must shower and replace outside clothes with sterile overalls, masks, hats, etc., before working with the animals. It is, however, a wise precaution on receipt of rodents to carry out some basic microbiological tests and other health screens. This ensures that batches of animals are free from diseases or abnormalities which might affect the course of the study or infect other animals.

On arrival in the facility animals should be marked, for identification purposes, and allowed to acclimatise to their new environment and diet. This acclimatisation period should last for at least two weeks before the commencement of a study, during which time various pretest measurements can be made. As it is usually considered desirable to commence toxicity studies in young rodents, six or seven weeks old, the animals should be about four weeks old and fully weaned on arrival. They should be weighed on several occasions before the study begins, and those failing to gain weight at a normal rate (or outside a normal weight range for their age) should be rejected. Detailed clinical observations should also be carried out on each animal before test. Those with obvious abnormalities, injuries or lesions should be rejected as should females where the vagina has failed to open or males with

undescended or abnormally sized testes. Before rats are allocated to groups, basic haematological and blood chemistry
tests (described in the next chapter) should be performed and
any animals with parameters outside the normal ranges should
be rejected. As a result of these selection procedures a number
of animals are invariably rejected. Therefore, to ensure that
adequate numbers are available for the study, it is advisable to
'over order' animals; an excess of about 10% is usually adequate. Having selected a suitable population of animals, they
should be randomly allocated to treatment groups using
random order tables or, more often nowadays, by a computer
generating random numbers.

At this stage the animals can be caged ready for dosing
although dosing should not commence until they have had a
few days to settle in their new environment. This is particularly
important if animals are being group housed for the first time
or if new individuals are introduced into a cage. Such 'social
problems' can be overcome by caging animals individually.
While this is obviously necessary in reproductive studies it is
often a disadvantage in most other types of long term studies
since singly housed animals are usually more aggressive and
can be more difficult to handle and dose. Another disadvantage of single caging is that it uses more space in an animal
facility and is more labour intensive. Although in some strains
of mice the males may fight when put together, this problem
can be minimised if they are group housed when they are very
young. Group caging does, however, have the drawback that,
in the event of a death, cannibalism can occur with the resulting loss of material for pathological examination. This problem can usually be avoided by very frequent inspection of the
animals, which is in any case good practice in toxicology
studies.

Housing environment has been observed to influence the
toxicity of some drugs. For example, for central nervous
stimulants such as amphetamines the LD_{50} in mice housed

singly is approximately one tenth of that for gang-housed (groups of ten) animals. Conversely with reserpine, a catecholamine depleting agent, the dose required to produce deaths in group-housed rats is lower than for single animals.

Decisions on single v. group housing must be made on a case to case basis with many factors influencing the choice, e.g. does the drug induce aggressive behaviour, space limitations, etc. Irrespective of the final selection, there are comprehensive recommendations on cage size and animal house environment which are listed in a publication by the Royal Society and the Universities Federation for Animal Welfare (1987).

Identification of animals within a group can be accomplished by permanently marking the animals. This has traditionally been by means of ear punching using a code which enables up to 9999 animals to be individually identified (an illustration of such a coding procedure is shown in Figure 3.1). Nowadays hypographs are available which enable tattooing of the tails of rats and mice, with individual numbers and bar codes respectively. Suitable models for such marking include those manufactured by Hypograph Systems, Langley Machine Tools, Barnsley, Yorkshire, and by Animal Identification and Marking Systems (AIMS) Inc., Piscataway, New Jersey, 08854, USA.

In cages containing several animals it is often useful to mark their fur to provide a quick and reliable way of identifying individuals during dosing and clinical observations. Such markings can be achieved quite simply in albino strains by using hair dyes such as Inecto (manufactured by Rapidol Ltd, London W7 2PP). The Institute of Animal Technicians' *IAT Manual of Laboratory Practice and Techniques* (1978) suggests the use of various histological dyes to give coloured markings, e.g.:

Yellow: saturated picric acid or chrysoidin.
Red: fuchsin

Violet: methyl violet (gentian violet)
Green: brilliant green, ethyl green or malachite green
Blue: trypan blue

Cages should also be clearly marked, with an individual label for each animal. If there are several animals in a cage it is best to provide labels with corners cut out to identify the individual animals as shown in the diagram (Figure 3.2). This enables individual labels to be removed when an animal dies,

Fig. 3.1 *Identification of individual animals by means of ear punching.*

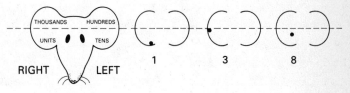

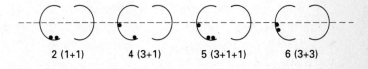

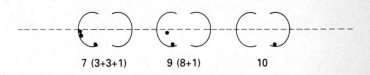

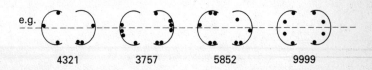

while still identifying the survivors. It is also usual to colour-code cage labels according to the study dose group. The same 'group colour coding' can be used for the stationery, etc., used for data recording in the study, thereby minimising the chance of confusion.

Cages should be arranged on racks in such a way as to distribute animals from different groups evenly in the various positions in the rack and around the animal room. This prevents variability due to minor differences in ventilation, heat, light levels, etc. It has been shown that even with a well controlled environment other factors such as animal placement may have an effect on the development of tumours. Such a response has been described by Young (1987) who reported that, in an oncogenicity study with eugenol, increases in hepatic tumours occurred in the first five of ten contiguous cages housing the low dose male animals. He concluded that some local room effect may have been responsible for the increase in tumours in the five affected cages. To overcome such an 'animal placement effect' it has been suggested that the cages housing animals should be randomly allocated in the cage racks and not organised contiguously, i.e. with the controls on

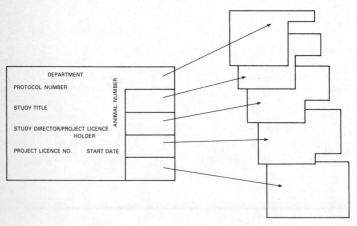

Fig. 3.2 *Use of labels to identify individual animals.*

the top, low dose next rack down followed by the middle and high doses. Random allocation of cages to rack positions is often used for large studies although it can make the task of dosing more difficult and confusing which may lead to inadvertent omission of a cage of animals from a scheduled dosing. We have found the system of diagonal arrangement more convenient for technical staff. With colour coding it is easy to see which cage should be dosed next and an even allocation of groups to the various rack and room positions is achieved.

Lighting cycles in animal rooms are usually 12 hours on and 12 hours off, switching on early in the morning. It must be remembered that rats and mice are essentially nocturnal by habit, so the animals will be active during the night. Behaviour during the day will be influenced by the fact that this is their normal sleeping period and thus locomotion, etc., will be minimal. Lighting levels used for rodent rooms are often too high, especially for albino strains, and retinal damage may result. With overhead lighting at levels which will be convenient for staff working in the room, the cages on the top tier of a rack should be protected by opaque or semi-opaque covers. Animal rooms should be maintained at a standard temperature, e.g. 21 ± 2 °C, and a controlled relative humidity in the range of 40–60%.

Standard diets for laboratory rodents are available from several major manufacturers. They can be provided sterilised for SPF facilities, and with a full chemical analysis. There is now some pressure to modify these diets to reduce the fat, protein, calcium and phosphate contents and to increase the magnesium levels. Many toxicologists are now convinced that laboratory rodents are overfed and obese, which may be responsible for reducing survival and increasing the risk of cancers of various types in these animals. Also a mineral imbalance in the diet may cause nephrocalcinosis in rats.

It has been demonstrated that dietary restriction can enhance survival and reduce tumour incidences (Tucker 1979,

Conybeare 1980). Dietary restriction can be achieved in a variety of ways, e.g. using only a limited amount of food per cage instead of the usual *ad libitum* regime, or by a time restriction where the food is available *ad libitum* for only a limited period (Conybeare 1987). Such time-restricted fed rats achieve normal body size, as indicated by brain weight and bone length, but other organs concerned with metabolism and excretion (liver and kidneys) are lighter than in corresponding *ad libitum* fed rats. Such animals do not suffer from premature ageing or high incidences of endocrine diseases. This latter effect can be seen in over nourished *ad libitum* fed female rats which often develop oestrous cycle irregularities starting at about six months of age, whereas time-restricted rats continue to cycle regularly until they are over 12 months old. Another consequence of *ad libitum* feeding is that animals will feed throughout the night and will have full stomachs in the morning when they are dosed, which can make gavage (dosing through a tube into the stomach) more difficult. At present, however, the attitude of regulatory authorities to studies conducted under diet-restricted regimes is not yet clear.

In order to control microbial contamination of the drinking water within a rodent unit the water system should be periodically flushed, at approximately six month intervals, with chlorinated water. The system, having been chlorinated for a minimum of eight hours, should then be drained and flushed several times with fresh water to remove all chlorine. As a usual procedure all drinking water is filter-sterilised to ensure animals are not exposed to pathogenic agents which may confuse toxicological evaluation.

3.2 OTHER RODENTS

Much of what has been said about rats and mice applies in principle to other rodents such as Syrian hamsters, gerbils and guinea pigs which are all readily available from commercial

breeders. Apart from the use of guinea pigs in topical toxicology testing (local administration, e.g. into skin, rectum,
eye), these other rodents are seldom used in toxicity studies
and are not available to the same standards as are rats and
mice. SPF strains are seldom bred at present, and microbiological and other health screen checks are therefore more important if these species are used.

3.3 RABBITS

Rabbits are used for teratology studies and for some short
term skin and eye studies. Several strains are widely available
(New Zealand White, Dutch, etc.), with some now beginning
to become available at SPF standards. They are generally
housed singly in metal or plastic cages.

3.4 DOGS

Dogs are the most widely used non-rodent species for toxicology studies, with pure bred beagles being the strain almost
universally used. There are several approved suppliers of Beagle in the UK and they can also be imported from the USA and
elsewhere. In the latter case they must be housed under quarantine conditions in facilities approved by the Ministry of
Agriculture, Fisheries and Food. Suitably designed toxicology
facilities, which obtain such approval, can conduct toxicology
studies with animals while they are being held in the quarantine facility. Irrespective of whether dogs are in quarantine or
not, they should be housed under precise environmental conditions. In most laboratories temperature is maintained at
about 18 ± 2 °C with air changes occurring between 12 to 15
times an hour. The water should be filtered with a 10 μm
preflow filter situated between the mains and the circulating
water tank with 0.2 μm filters being placed in the circulating
water system.

It is difficult to control the relative humidity rigidly in a dog area since it can depend upon a variety of factors, e.g. number of animals in the room, routine washing of bays, excess amounts of water produced as dogs use automatic drinkers. It is still usual, however, to measure relative humidity at least once a day, usually when the bays are clean and dry, and aim to maintain it at about $55 \pm 10\%$.

To check on the functioning of heating, ventilation etc., each parameter should be monitored automatically at set time points. The lighting levels, nominally 600 lux, should also be automatically controlled giving a 12 hour light (6 a.m. to 6 p.m.) and 12 hour dark (6 p.m. to 6 a.m.) cycle. Although outside the UK dogs are often kept in cages, in the UK they are expected to have pens with a minimum floor space of $4.5-8$ m^2 if housed singly or $10-40$ m^2 if multiple housing is used (dependent on size of dog). The minimum pen height should be 2 m.

On receipt of dogs, a thorough veterinary examination of each animal should be undertaken as well as the usual microbiological and other health checks. As a precaution it is usual to treat the animals with an anthelmintic before undertaking any pretest analysis of blood and urine samples. As with rodents, more animals should be purchased than will be needed for the test so as to allow for rejection of animals considered unsuitable for toxicology testing. Reasons for rejection can include electrocardiographic and ophthalmological abnormalities, clinical chemistry, haematology or urinalysis parameters outside the normal range and animals outside the normal weight range for their age and sex. Dogs are usually identified numerically by breeders by having numbers tattooed on the inside of their ear pinnae. After allocation to dose group, identification of individual dogs can be by a tattoo in the other ear pinna or by a collar and tag system.

Acclimatisation of dogs should preferably be for a minimum of four weeks before dosing is due to begin, during which time at least two blood samples should be taken for clinical chemistry and haematology examinations. Such measurements help in animal selection and establish a baseline and background normal data before the commencement of treatment.

Interanimal variation should be kept to a minimum in a toxicity study by ensuring that the animals on a particular study are within a narrow weight and age range. Such weight and age uniformity is necessary because toxicities have sometimes been found to be greater in animals of high body weight. Usually, for economic reasons, dogs from commercial suppliers are relatively young, but ideally they should be at least five or six months old before dosing begins and gaining weight at a normal rate.

Allocation of dogs to treatment groups is a conscious selective process (unlike rodents which is done on a purely random basis). The reason for this is because dogs supplied for experimental purposes are not so genetically or phenotypically uniform as rodents. Thus in a group of Beagles from a commercial supplier there will usually be dogs of different weights, ages (usually one to two months) and a variety of sibling relationships. It is therefore necessary to ensure that any one particular group does not contain several siblings or have an average body weight, clinical chemistry parameter, etc. which is significantly different from the other groups. It is therefore usual to assign dogs with particularly high or low values for any parameters evenly throughout the groups in order to avoid bias.

Standard pelleted diets for dogs are available from several commercial companies. They are usually provided with a full analysis of content, and dogs on such diets require no other food supplements. Some form of dietary restriction is needed, particularly for long term studies, in order to avoid the problem of obesity. Such restriction can be in the length of time the food is presented but is more usually in the quantity offered.

3.5 NON-HUMAN PRIMATES

Simians of various types are widely used in toxicology studies and while a few are bred in captivity, most are still captured in the wild. A major problem with wild caught monkeys is the uncertainty of their health status, which may pose health risks to staff and make interpretation of toxicological effects difficult, especially in the presence of 'background' pathological lesions.

Apart from some UK bred marmosets virtually all primates are imported. Although the animals are usually given a thorough health check and 'conditioning acclimatisation' by the importers before being supplied to laboratories, they should still be given a very thorough examination before being used in studies. All facilities using primates need to be approved for quarantine purposes, since fully quarantined imported primates are not readily available.

The principles described for dogs, in terms of pretest examinations, selection and allocation to groups, are equally applicable to primates. Identification is usually by tattooing on the arm or body of the animal. Caging and housing requirements for primates are described in the guidelines from the Royal Society and UFAW (1987).

Unfortunately the ages of wild caught monkeys are not precisely known, although some guesses can be made based on size and appearance. With the larger species such as baboons, rhesus or cynomolgus monkeys, the animals supplied by the importers are in a weight range in which the females may be sexually mature but the males are not. Since it is desirable in toxicity studies to use sexually mature animals this can sometimes prove to be something of a problem.

Commercially available pelleted diets for primates, while perfectly satisfactory in nutritional terms, are usually supplemented by fresh fruit. Dietary restriction is not normally necessary.

3.6 OTHER NON-RODENTS

Although other non-rodent species such as ferrets, cats, sheep and pigs are seldom used for toxicity studies they are available and some of the same principles for selection and allocation which have been discussed early will apply. Because of the relatively small group size and the non-homogeneity of their genetics and health status such animals would be allocated to groups in a non-random way after the assessment of pretest data. Since it is impracticable in this book (and indeed not its purpose) to describe all the various laboratory animals which can be used in toxicological studies, the reader is therefore directed for further information to *The UFAW Handbook on the Care and Management of Laboratory Animals* (1987).

REFERENCES

Conybeare, G. (1980) Effects of quality and quantity of diet on survival and tumour incidence in outbred swiss mice. *Food and Cosmetic Toxicology*, **18**, 65–75.

Conybeare, G. (1987) Modulating factors – challenges to experimental design. In *Carcinogenicity: The Design, Analysis and Interpretation of Long Term Animal Studies*. Springer-Verlag, New York.

Institute of Animal Technicians (1978) *IAT Manual of Laboratory Practice & Techniques*. Granada, St Albans.

Poole, T. (1987) ed. *The UFAW Handbook on the Care and Management of Laboratory Animals*. Longman, London.

Riley, V. (1975) Tumours: alterations of incidence an apparent function of stress. *Science*, **189**, 465–567.

The Royal Society & UFAW (1987) *Guidelines on the Care of Laboratory Animals and their Use for Scientific Purposes*. The Royal Society and the Universities Federation for Animal Welfare, London.

Tucker, M. (1979) Effect of long term diet restriction on tumours in rodents. *International Journal of Cancer*, **23**, 803–12.

Young, S. S. (1987) Are there local room effects on hepatic tumours in male mice? An examination of the NTP eugenol study. *Fundamental and Applied Toxicology*, **1**, 1–4.

4

Standard studies in animals

In the process of developing a new drug, a series of animal toxicology tests of increasing duration are conducted. It might appear to make good scientific sense to run the studies of increasing duration sequentially so as to utilise the findings of a shorter study before designing a subsequent longer study. However, since patent protection of a drug is of limited duration (15–20 years) it is commercially important to exploit as much of the patent protected period as possible. Since, with many drugs, oncogenicity studies lasting two years or more will be necessary before marketing can begin, some telescoping of the sequence of toxicity tests is usually considered desirable. A calculation must therefore be made balancing the cost of studies against the likelihood of a successful commercial drug development.

4.1 GENERAL PRINCIPLES

Before describing the methods used in standard animal studies it may be useful at this point to discuss some general issues associated with toxicity studies, e.g. selection of species, group size, dose selection (more pertinent to subchronic and chronic studies) and route of administration.

Species selection

The choice of species for use in toxicological studies is a compromise between what is required or desired, and what is practical. Naturally, when performing any toxicity study it would be ideal to use a species with the same metabolism, pharmacokinetics and target organ susceptibility as man. It has been argued that regulators should pay more attention to pharmacokinetics and mechanisms of toxicity rather than relying on a regulatory checklist (Clarke *et al.* 1985). A major problem is that before it is possible to give a compound to man, in order to learn something of human absorption, distribution, metabolism, excretion, kinetics and pharmacodynamics, it is essential to have animal toxicity data (usually up to 30 days, in two species; see Chapter 2). It is therefore possible that a large resource effort has already gone into performing toxicity studies in two species, as dictated by regulatory authorities, which may have different metabolic and pharmacokinetic profiles from that of man. It is expensive in terms of resources and time (patent protection) to begin searching for species having similar metabolic profiles to that of man, and then having to repeat all the early toxicology studies in these species. However, if in human studies a major metabolite is produced which is not found in any of the available species, it may then become necessary to synthesise the metabolite and study its toxicity in a common laboratory species.

Other problems such as availability of a suitable standardised diet, housing facilities (especially for long term studies) and psychology of the animal (some species may become very stressed over long periods of dosing), make species selection outside the conventional range of laboratory animals very difficult. For these and other reasons, such as continuity and availability of historical data, it is likely that the common laboratory species, e.g. rats, mice, dogs and monkeys, will continue to be used, with allowances being made for differences in metabolism. Naturally if a less commonly used but commercially available species, such as the pig, was found to be metabolically much closer to man than dogs or primates are, it is likely that studies – probably up to 12 months' duration – could and would be conducted in such a species.

For oncogenicity studies the established practice is to conduct lifetime studies in rats and mice, i.e. 24 and 18–21 months respectively. Sometimes, however, it may be necessary to consider using another rodent such as the hamster or in rare cases a non-rodent species, e.g. the dog or a primate. The latter course of action is extremely costly and may require up to seven years' experimentation. Although such studies have been undertaken with contraceptive steroids, in general, for practical reasons, the rat and mouse remain the animals of choice for carcinogenicity studies.

Group sizes

The minimum number of animals in treatment groups is dictated by the regulatory authorities (see Chapter 2) and is discussed for individual studies later in this chapter. However, in a complex situation, perhaps studying an effect which occurs only rarely but is of great biological importance, it may be advisable to increase the numbers of experimental animals. Once again the desire for accuracy and sensitivity must be weighed against certain practicalities not least being that for ethical reasons, as few animals as possible are used in toxicological studies.

Very often 'satellite' groups of animals are included to provide extra material for investigative studies or additional haematological or clinical chemistry parameters. These animals do not form part of the 'mainstream' study and are not usually used for histological evaluation. If mice are used as a toxicology species their size often makes it necessary to use satellite groups for routine clinical chemistry and haematology measurements.

In oncogenicity studies, although group sizes of 50 males and 50 females per treatment are recommended, it is not unusual to use groups larger than these. Such studies may also include two control groups, each containing similar numbers of animals to the treatment group. The reason for the increased numbers of controls is to examine the frequency of statistically significant differences between two 'identical' concurrent control groups. Such findings are important if it can be demonstrated that an observed difference between control and treatment groups may not be related to treatment but may simply have occurred by chance

Route of administration

As a general rule drug administration in animals must include the route to be used in human dosing, as well, usually, as the oral route and frequently the intravenous route. Thus oral administration is employed for most compounds even though it might not be the intended route for use in humans. One reason for this is the need to predict the consequences of accidental or suicidal ingestion. The intravenous route (see Chapter 7) is useful in providing a measure of acute toxicity unaffected by such factors as metabolism or absorption.

The problem imposed by the regulatory need to use the proposed clinical route of exposure in toxicology studies, together with the practical difficulties in using routes other than oral in long term studies, is frequently resolved by employing the clinical route for short term studies, to evaluate local toxic effects, and oral dosing for chronic studies. When using this approach to examine both local and systemic effects it is necessary to ensure that drug blood levels achieved by the oral route are at least as high as those achieved by topical dosing. The only significant exception from this is dosing by inhalation, by which long term and oncogenicity studies are conventionally conducted.

For most drugs it is the oral route of administration which is most commonly used for long term studies, the method of dosing being dependent upon the circumstances of the study. When rodents are used the drug is usually given as a suspension or solution by gavage, although for longer term studies it can be incorporated into the diet or even the drinking water, provided that the drug is stable and can be distributed homogeneously. While this latter method may be the most popular for compounds given at very low doses and for compounds other than drugs, e.g. herbicides, pesticides and food additives, it has a number of drawbacks for pharmaceuticals. First it must be recognised that the dosing is not very accurate. Often animals are not caged singly so that the best estimate of the amount they have eaten can only be the mean for the cage. Another problem is that very few drugs have no taste – pleasant or unpleasant. The taste of a drug can influence the amount of food consumed and it is well established that the dietary intake of an animal plays an important role in its health, survival and rate of tumour incidence. Also changes in food intake can affect certain laboratory measurements, e.g. clinical chemistry parameters (see Chapter 5), thereby possibly giving a false indication of a toxicological response. It is also not uncommon to find that rodents learn to separate the drug from their diet and avoid eating it, although this problem can sometimes be resolved by formulating the drug/diet mixture into pellets. Another problem is that drug-contaminated dust can be generated from such drug/diet mixtures, which may be inhaled or ingested by other animals, including controls, present in the facility.

Dosing by gavage also has its problems. It requires skill on the part of the technical staff to avoid dosing accidents, and is more costly in labour – especially during weekends and bank holidays. High doses of hypertonic, sometimes irritant, solutions or suspensions at high or low pH can also cause accidental toxicity in the respiratory tract by reflux from the oesophagus to the trachea. Such problems can, however, be minimised by use of suitable catheters (Conybeare & Leslie 1988).

Dosing by gavage has been described as being analogous to giving repeated acute doses, whereas giving the drug in the diet provides a more sustained exposure. It has been suggested that this could be the reason why observed toxicity may vary considerably between dietary administration and gavage. Neither method, however, can be said to be closer to the clinical dosing situation, where dosing may be from one to four times daily. For non-rodents, while the dietary and gavage methods of dosing can be used, it is more common to give the compound in capsules or in formulated tablets, often containing similar excipients to the clinical formulation.

Dose selection

The usual way of conducting a toxicology study, as advocated by the regulatory agencies, is to use a three dose (high, medium and low) design with a negative (placebo or vehicle dosed) control group. The basic idea is to assess any target organ toxicity directly and to determine a threshold effect. By characterising the toxic response over an experimental dose range it is possible not only to compare the relative toxicity of the test agent in different species but also to determine a 'no observed effect level' (NOEL), i.e. the level of the test agent which can be administered to an experimental animal without producing toxic effects. Determination of the NOEL is important in order to ensure that the dose of drug which produces the desired pharmacological response is lower than that responsible for causing any toxicological side effects.

Selection of the 'low' dose is usually based on giving a low multiple (two to five times) of the probable clinical dose. In early stages of drug development, prior to clinical trials, the ED_{50} in animal models provides the best available estimate of this dose. In subsequent toxicology studies, however, the likely therapeutic dose range should be better established and used to ensure that no substantial toxic effects occur at treatment levels in low multiples of this range.

Selection of the high dose level can be a particular problem. There is general agreement that the top dose should produce toxic effects. Indeed, in acute, subacute and subchronic studies, when attempting to identify target organs of toxicity, the dose may be so high as to cause death. In chronic studies, however, although the top dose is expected to produce some toxicity, thereby demonstrating the sensitivity of the experimental animals, the treatment should not cause appreciable animal deaths or reduce their lifespan, since we do not want the animals to die before they can demonstrate any specific findings or have the chance to develop tumours.

Most toxicologists would accept a dose that causes a 10% reduction in body weight gain as one that causes minimum toxicity. Many would consider doses of 100 or more times the proposed clinical dose as being of little value in predictive toxicology. This is especially true of compounds with a potential to affect 'normal' physiological functions markedly, or to cause irritation. Although such responses may cause oncogenesis, this is not necessarily predictive of such an effect at therapeutic doses in man. It therefore follows that it is essential, especially in oncogenicity studies, to monitor carefully the physiological status of the animals throughout these long term studies.

It is possible, however, that we may be confronted with a compound having very low toxicity and of course there are physical limitations to the amount of drug which can be administered, even by the oral route. It is difficult and probably meaningless to administer doses higher than 1 or 2 g/kg daily, and for drugs administered in the diet 5% is generally considered to be a sensible maximum. Above these dose levels the potential for adverse effects caused by volume, tonicity, pH, interference with absorption of dietary constituents, catharsis or effects on food or water intake make realistic interpretation of data impossible. It is also pointless to dose at levels beyond which further absorption does not occur, as any additional compound beyond this level is, in terms of bioavailability, merely wasted.

Another problem which can occur is that the pharmacological activity of the test drug may preclude giving very high doses. An example of this could be certain types of centrally acting drugs. For instance, if an oral hypnotic was given at a multiple of its effective dose it could cause such prolonged sleep that the animals would not wake up sufficiently to eat. Thus, while in general it may be desirable to treat at toxic levels, in certain instances this may not be possible.

Another point to remember, especially with chronic studies, is that the ability of animals to absorb, distribute, metabolise and excrete drugs can vary quite considerably with age. While a young animal may be able to cope quite well with drug treatment, as it gets older its metabolic capabilities may start to change, possibly resulting in drug accumulation and tissue overload. Both of these effects can result in severe toxicological problems for the 'geriatric' rat and mouse. (This effect is especially relevant in oncogenicity studies.) Because of these responses it is possible that dose level selection will change depending on the type of study which is being conducted. Thus the top dose in a rat oncogenicity study may have to be considerably lower than the top dose used in a 30 day study.

It is usual to include at least one intermediate dose group in most toxicology studies. However, when studying drugs with idiosyncratic toxic reactions, it may be advisable to have two or even more. Such intermediate dose groups, as well as being a regulatory requirement for most toxicology studies, can provide the toxicologist with important information about dose-related effects. Such information may prove vital if, for example, a toxic effect at the high dose level is due to metabolic, pharmacological or physiological overload. The absence of the response at the low and intermediate doses, especially in the case of oncogenicity, may save a valuable drug from being wrongly considered unsafe at therapeutic doses.

An important point about dose selection is that, while animals may receive 'similar doses' in terms of mg drug/kg body weight, pharmacokinetic analysis may demonstrate marked differences in terms of drug blood concentration and/or drug retention time. Thus different species given similar doses may show very different drug concentrations in blood or tissue. Such responses are extremely important since there is little point performing toxicity studies in a species which does not absorb the test agent. By knowing the pharmacokinetics of the drug in several species it is possible to compare effects not only in terms of dose given (mg/kg body weight) but also in terms of blood concentration.

4.2 STANDARD STUDIES

Single dose (acute) toxicity

These studies serve to establish a lethal dose range and provide prompt warning if a highly toxic compound is being dealt with. They also provide information on limiting toxicity due to pharmacological effects and on target organ toxicity, and assist in establishing the maximum dose to be used in subsequent subacute studies. This latter information is important for making some prediction of the amount of chemical required for future toxicological studies. (Such a prediction of compound requirements is essential in drug development since it is obviously necessary to have an adequate supply of chemical available when required.)

Because quite precise measures of acute toxicity can be made, such studies can provide important data for the selection of candidate compounds for future development. This use of acute toxicity data to make a 'go' or 'no-go' decision about development is, however, usually of more importance with compounds other than drugs. This is because in certain industries it is possible to have many candidate compounds, all of which may possess the desired chemical or biological property. In the pharmaceutical industry we are unlikely to be in the fortunate position of having many alternative compounds with similarly desirable pharmacological profiles. Finally there is a regulatory need in most countries to provide acute toxicity data. Fortunately most countries (with the notable exception of Japan) no longer require a formal accurate LD_{50}, and will accept acute toxicity levels established by hierarchical methods like that described recently by a working party set up by the British Toxicological Society (Anon, 1984).

In the rodent three types of acute toxicity study may be performed. (Most toxicologists would consider it unethical to carry out acute studies in non-rodents.) As a first step it is usual to establish a maximum tolerated dose (i.e. highest dose not causing death or life-threatening toxicity) and the minimum lethal dose. Second, there are single dose studies to establish target organ toxicity, and third, there is the determination of precise LD_{50} or median lethal dose. This last study may be required in certain countries for a clinical trials certificate or even a product licence. It is, however, unusual to perform an LD_{50} estimation until the compound is well into development and it is known that clinical trials are to be conducted in those countries requiring this information.

In the first type of study two species, normally mice and rats, are dosed both orally and by intravenous (IV) injection with the test agent. (Naturally if the proposed route for clinical use is neither IV nor oral, this route must also be used in acute studies.) The most efficient method, in terms of compound and animal use, is to treat one male and one female animal with a high dose of the material, e.g. 100 mg/kg for IV injection or 1000 mg/kg for oral dosing. These animals are then observed for about an hour; if the animals die, a second pair are treated with half the dose. The procedure is repeated, i.e. the animals are examined for one hour and the dose continues to be halved, until a treatment is reached at which the animals will survive. Alternatively, if the animals survive the hour a second pair receive double the dose. (It is not usual to go beyond an oral dose of 2000 mg/kg.) If such a treatment is tolerated a second pair of animals is dosed at this concentration for confirmation. For greater precision doses may of course be increased or reduced by less than a factor of two. The surviving animals are then left untreated for 14 days, during which time they are observed daily for the development of clinical symptoms. On day 15 they are killed and subjected to a thorough examination *post mortem*.

In the second type of acute study, to determine target tissue and an approximate median lethal dose, four to six randomised groups of animals (three to five males and three to five females per group) receive a single treatment of test agent. The doses used are in the 'lethal dose range', as determined from the study described previously. Rats and mice are again used, with the chemical being dosed both orally and IV. The animals are observed daily for 14 days with a full necropsy being performed on all decedents and survivors. All abnormal tissues plus major organs such as heart, liver, kidney, lungs are usually taken for histological examination.

To establish a precise LD_{50} at least four dose levels are used, with 5–10 males plus 5–10 females per treatment group. The animals are given a single dose and, at the end of the 14 day observation period, the survivors are subjected to necropsy with the major organs and abnormal tissues being taken for histology. The LD_{50} and its 95% confidence limits are calculated from the lethality data using probit analysis.

Dose ranging and sighting studies (repeat dosing)

The purpose of the dose ranging study is to investigate the toxicity of repeated doses of an agent at dose levels ranging from a very low multiple of the therapeutic effective dose (i.e. somewhere in the range of the ED_{50} for the species to be used, or the anticipated human therapeutic dose range), up to doses approaching the maximum non-lethal dose (this latter dose having been established in rodent acute toxicity studies).

Approaches to dose ranging studies differ depending on the species being used. In rodents it is usual to perform a seven to ten day repeated dose study (dose ranging study). The procedures used can vary but usually involve exposing experimental animals (group size two to five animals/sex/group) to various concentrations of the drug, i.e. from the maximal non-lethal dose found from acute studies down to the 'pharmacological' dose. Clinical chemistry and haematological parameters are measured at the beginning (48 hours after first dose) and end of the study, with full histopathology being undertaken on all animals (including decedents). The data from such studies provide further information on the toxic and lethal potential of the drug and assist in dose selection for subsequent subchronic studies.

In non-rodent dose ranging studies the number of animals used may be as low as one per sex with the same pair of animals receiving increasing dose levels in a stepwise sequence until a maximum tolerated dose is reached. In practice one male and one female dog (or monkey) are usually treated on two consecutive days with a preselected dose of the compound, e.g. five times the ED_{50}. On the third day the dose is raised and the animals are treated at this level for a further two days. If the animals tolerate this treatment, the dose will be raised on the fifth, seventh, ninth day, etc. until the highest tolerated dose is achieved. This is called an ascending phase maximum tolerated dose (MTD) study. In order to maximise the opportunities for detecting target organs or systems, and to provide a basis for helping to assess the doses for subsequent studies, the animals should be maintained on this dose for up to a week. After this they are subjected to a full necropsy with the major organs, i.e. heart, liver, kidney, lungs, together with abnormal tissues being taken for histological examination.

During the course of the study a full panoply of clinical observations are performed with blood samples being taken every alternate day, i.e. on days when dose levels are increased. In these stepwise (incremental) studies the dose may be doubled on alternate days if a large ratio between ED_{50} and the toxic dose is anticipated. The increments would, however, be smaller when lower therapeutic ratios are expected.

It must be remembered that it is *not* the objective of dose ranging studies to determine a lethal effect, and indeed when overt adverse effects are encountered the dose is normally reduced. With certain types of adverse response, particularly limiting pharmacological effects, tolerance may develop and it may be possible to raise the dose again to the previous high level after a day or two of treatment at the lower concentration.

Following the ascending phase study it is usual to perform a constant phase (fixed dosage) MTD study in which groups of dogs/monkeys (one or two animals/sex/group) are dosed daily (seven to ten days) with concentrations of chemical covering multiples of the pharmacological dose up to the MTD (this having been determined by the ascending phase MTD study). Clinical chemistry and haematological parameters are usually measured at the beginning and end of the study, with all animals being subjected to full histopathology. These studies not only provide valuable information on the identity of possible target organs and assist in rational dose selection for subchronic studies, but also provide data on the response of 'naïve' animals to high doses of the drug. In the ascending phase it is often the case that dogs develop a 'tolerance' to the drug as a result of being exposed to continually increasing doses and thus can eventually survive relatively high doses. On the other hand 'naïve' animals, having no previous exposure, often demonstrate toxic susceptibility at lower concentrations. Since in the subchronic doses 'naïve' dogs are used, it is important to ensure that the high dose animals will tolerate the selected dose and not die during early dosing, thereby possibly invalidating the study.

Although these dose-ranging, or sighting, studies are not usually required for regulatory purposes, they may be used in applications to Japan in place of a non-rodent LD_{50} study. The need for high standards in these non-regulatory studies is obvious, since the data produced are fundamental to the design of subsequent toxicological investigations. As such acute studies employ higher doses than will be used in subsequent studies, they may also prove to be important in establishing target organ toxicity.

Two week and four week repeat dosing

These studies have two purposes. First, they are formal studies which may be used to satisfy Regulatory Requirements prior to initiating the first human administration of a compound. Data from four-week toxicity studies will usually support clinical trials allowing repeat dosing to be given over several days (the actual duration of treatment being dependent upon the country in which the clinical trials are undertaken and the nature of the drug), while two week data will usually only permit one day of human administration. Since the additional effort of dosing for two extra weeks is such a small proportion of the work, it is often considered not worth while to run two-week studies unless the compound is in very short supply. The second purpose of these subchronic investigations is to help set doses for subsequent longer term studies. In order to identify target organs or systems at risk it is obviously necessary to achieve the highest possible tolerated doses. This may necessitate adjustment of doses during the course of the studies if it becomes obvious that they are too high or too low. This flexible approach should produce predictive adverse effects, while ensuring the survival of sufficient animals to the end of test for full histopathological evaluation. This flexibility will also be evidenced by the interpolation of extra blood chemistry, haematological or other investigative procedures to aid interpretation of drug-induced effects. Subchronic studies are therefore likely to have a series of protocol amendments.

For these studies the group size is usually about ten of each sex per group for rodents and three or four of each sex per group for dogs or primates. The duration of a 'four'-week study is frequently 30 days rather than 28, since the former period is required by some regulatory guidelines. As stated previously, all such studies suffer in their design from the handicap of not knowing to what extent they are relevant to the human situation, since metabolic and pharmacokinetic data from humans are lacking at this stage of the drug development process.

Long term dosing (chronic) toxicity studies

The boundary between subchronic and chronic studies is not too clear and perhaps not important either. While two or four week studies are usually classified as subchronic, those conducted for 6 or 12 months are classified as chronic. This is because, for mice and rats, the length of dosing represents a substantial proportion of their $2^{1}/_{2}$ or $3^{1}/_{2}$ year life spans. Naturally, for dogs or primates, a 6 month or even a 12 month study covers only a small part of the expected life span.

Chronic toxicity studies in two species (rodent and nonrodent, in practice usually rat and dog respectively) are required for both long term clinical trials and marketing approval (see Chapter 2 and Appendix 1). In a number of countries the duration of a chronic toxicity study depends upon the type of drug and the proposed duration of treatment in man. Other authorities, however, are not so flexible. The Canadian regulators, for instance, require 18 month studies for drugs to be prescribed for relatively short periods. In Australia, Japan and the USA the authorities accept 12 month studies for drugs to be used for long term treatment (possibly up to 6 months), while in Europe 6 month animal studies are all that are required to treat humans for a similar period of time. (For drugs to be given to humans over longer periods, oncogenicity studies are also required.)

There has been considerable debate recently about the scientific justification for the very long studies required by some countries. Frederick (1986) has argued the case for the Canadian guidelines requiring 18 month toxicology tests on the basis of 15 cases cited where adverse effects were (or could have been) missed in studies limited to 12 months or less. However, Lumley & Walker (1986), after reviewing data provided by 21 companies on 124 compounds, concluded that tests of longer than 6 months did not add to the overall safety evaluation of compounds' excluding the possibility of detecting carcinogens. Emmerson (1987) concluded that the data from a 12 month study was ample for a clinical development programme, while Roe (1986) pointed out that for most drugs oncogenicity studies would be conducted, and that these should obviate the need for very long term chronic toxicity studies.

Group sizes in chronic studies are usually larger than in subchronic studies. At least 20 of each sex per group being usual for rodents and 5 or 6 of each sex for non-rodents. Additional animals may be used in the control and highest dose groups. This is to provide evidence for 'reversibility' of any adverse effects by including a treatment-free 'recovery' period at the end of the treatment phase. Subchronic studies should have provided data on organ toxicity which, together with information on metabolism, absorption and pharmacokinetics, should enable the doses to be used in chronic studies to be set with some confidence, and therefore radical changes in dosage are less common. However, adverse effects may be detected which were not observed in studies with shorter exposure to the drug and additional investigations may well be required as the studies progress.

Blood and urine samples are always collected during these

studies. The precise timing and frequency of such sampling are dependent on a variety of factors, e.g. length of investigation, toxicological responses seen in earlier studies, etc. An example of sampling times over a 12 month study could be weeks 5, 13, 26 and 52.

Oncogenicity studies

The majority of regulatory authorities require that oncogenicity studies, in two species using three dose levels, be performed with all drugs destined to be taken regularly over a substantial period (i.e. longer than 6 months) or if repeated short periods of treatments are envisaged. Such oncogenicity studies are in addition to the chronic studies discussed previously.

The overwhelming majority of oncogenicity studies are conducted in rodents, although there are a few situations where oncogenicity studies in non-rodents may be deemed necessary (e.g. if the test agent has a hormonal action, as with oral contraceptives, rodents have given misleading impressions of carcinogenicity. The oncogenic potential of these compounds may therefore have to be examined in primates or dogs. Because of the longevity of these species, dosing may last for seven to ten years.) Two species are required and, while some studies, especially inhalation studies, have been conducted in the golden hamster (Syrian squirrel), most studies are carried out in rats and mice. It is considered necessary, in order to maximise the predictive value of oncogenicity studies, that exposure to the compound should be for the majority of the life span of the animal. In order to achieve this in a reasonable time only small short-lived animals are used. The optimum duration of such studies in rats and mice has been much debated. On the one hand, the longer the exposure the greater will be the opportunity for an oncogenic potential to become apparent. On the other hand, in very senile rodents some tumourous changes due to drug action may be impossible to detect because of the high background of 'geriatric' pathology.

Most commonly rats are dosed for 24 months and mice or hamsters for 18 or 21 months. This is acceptable to regulatory authorities, and most strains of laboratory rats or mice maintained in SPF conditions will remain in relatively good health throughout these periods. This is important since mortality due to causes other than tumours should be less than 50% after 24 months administration for rats, and after 18 months for mice and hamsters. The loss of animals due to autolysis, cannibalism, etc. should not exceed 10% in any treatment group. Accordingly, if the animals are emaciated or moribund they should be killed and subjected to necropsy. Sometimes studies are designed so that the duration depends on the survival rate encountered during the study. For example, the study may be terminated for each sex when only 25% of that sex are surviving in the medium dose group or in the control group. Such designs make planning the use of a facility or staff very difficult and the scheduling of necropsies on so many animals found dead or killed *in extremis* may also cause considerable logistic problems.

The group size employed in rodent oncogenicity studies is at least 50 of each sex and it is essential to select a top dose level which, although exhibiting minimal toxicity, should permit most animals to survive until the end of the dosing period. In such long studies it is vital that those staff carrying out the dosing should have sufficient skill and training to enable them to ensure that very few deaths are caused by misdosing. In studies where compounds are to be administered by diet, it must be remembered that survival, the incidence of many tumours and other pathological changes in rodents are influenced by the quantity and composition of the diet (Conybeare 1980, Roe 1983). Any effect of the drug on the amount of diet consumed could cause major treatment-related differences in the incidences of neoplasms and in the pathology of major organs such as the liver and kidneys.

It is most unusual to perform any laboratory investigations, e.g. clinical chemistry and haematology, during the course of an oncogenicity study. There are a number of reasons for this, e.g. size of the study (the resource effort required for bleeding many hundreds of animals, performing the measurements, collating results, etc., is very large), extra stress on the animals which may result in premature death or the development of non-treatment-related tumours (this may be of particular importance in the high dose animals where treatment may also be producing stressful effects), chronic rodent studies include laboratory measurements, etc. Despite these reasons a number of individuals, and now certain authorities, have queried whether valuable toxicological information is being lost by not performing extra measurements during the course of oncogenicity studies. It is possible that, in the future, regulatory authorities may require that various measurements, such as clinical chemistry and haematological parameters, are made during the course of these studies. Before such demands are made, it is to be hoped that advice on the suitability of such measurements in oncogenicity studies will be sought from educated sources, that full and frank discussions will take place and that wise judgement will prevail.

REFERENCES

Anon (1984) Special report: a new approach to the classification of substances and preparations on the basis of their acute toxicity. *Human Toxicology* 3, 85–92.

Clarke, A. J., Clarke, B., Eason, C. T. and Parke, D. V. (1985) An assessment of a toxicological incident in a drug development program and its implications. *Regulatory Toxicology and Pharmacology*, **5**, 109–119.

Conybeare, G. (1980) Effect of quality and quantity of diet on survival and tumour incidence in outbred swiss mice. *Food and Cosmetic Toxicology*, **18**, 65–70.

Conybeare, G. & Leslie, G. B (1988) An improved oral dosing technique for rats. *Journal of Pharmacological Methods* 19, 109–116.

Emmerson, J. L. (1987) Letter to Editor. *Fundamental and Applied Toxicology*, **8**, 134–8.

Frederick, G. L. (1986) The necessary minimal duration of final long-term toxicological tests of drugs. *Fundamental and Applied Toxicology*, **6**, 385–94.

Lumley, C. E. & Walker, S. R. (1986) A critical appraisal of the duration of chronic animal toxicity studies. *Regulatory Toxicology and Pharmacology*, **6**, 66–72.

Roe, F. J. C. (1983) Carcinogenicity testing in animals and alternatives. In *Toxicity Testing*, ed. M. Balls, R. J. Riddell, & A. N. Worden. Academic Press, New York. pp. 127–30.

Roe, F. J. C. (1986) How long should toxicity tests in rodents last? Editorial in *Human Toxicology*, **5**, 357–8.

5

Measurements and observations made in living animals

During the pretest phase, after animals have been allocated to treatment groups but before treatment begins, a series of tests and observations are usually undertaken. These are repeated at various intervals during the in-life phase of the study. They are designed to measure a variety of parameters which help to determine the health of the animal, and to indicate whether any adverse effects are being caused and if so, identify the time of their onset. Some of these observations are non-invasive and simple to perform, while others may use invasive techniques (such as regular sampling of body fluids), and require sophisticated and automated analytical equipment. The frequency of such monitoring depends to a large extent on the measurements being made, the nature and duration of the study, and the species under test.

5.1 OVERT BEHAVIOUR

Assessment of the behaviour of animals should be carried out 'informally' on a daily basis before dosing begins and throughout the study. More formal clinical observations, such as those described below, should be conducted before dosing begins and thereafter at regular intervals throughout the in-life phase.

Since these observations form the basis for any treatment-related changes in behaviour, they should be recorded in an appropriate log or record sheet, dated and initialled.

Before examining an animal its identification should be verified by reference to its ear marking, tattoo or collar tag. Rodents should be placed on a flat surface and their posture, movements and behaviour observed. They should then be gently felt from head to tail to check the condition of the fur and for damaged areas of skin, subcutaneous swellings or lumps (the size, shape and consistency, i.e. hard, soft, etc., of such lumps being measured), areas of tenderness, abdominal distension, etc. All natural orifices should be checked for discharges, blockages, blood, mucus, etc., and the eyes examined for dullness, dryness, discharges, opacities, pupil diameter, ptosis (drooping of upper eyelid), exophthalmos (abnormal prominence or protrusion of the eyeballs), etc. The colour and consistency of the faeces should be noted and any wetness or soiling of the perineum recorded. The mouth should be examined for excessive salivation, lumps, cuts, etc., and the teeth checked to see if they are straight, parallel, of equal length and correctly aligned. Any breathing abnormalities such as sneezing, wheezing, rattles or dyspnoea (difficult or laboured breathing) must be recorded. In acute studies more detail must be paid to behavioural changes and some attempt made to quantify these and record them in a format similar to that suggested by Irwin (1962).

5.2 BODY WEIGHT

Changes in body weight can be a sensitive and important monitor of the health of an animal. Loss of body weight, or failure to gain body weight at a normal rate, is frequently the first indication of the onset of an adverse effect. In some cases the dose at which this effect is seen may be considered to be the treatment limit for long term studies, i.e. treatment producing body weight loss of 10% may be considered to be a toxic dose, irrespective of whether or not it is accompanied by any other changes.

In acute and short term studies body weights should be recorded daily. In longer term studies weekly recordings should commence before the treatment period starts, so that animals which are failing to gain body weight at a normal rate can be rejected. Nowadays many on-line data recording systems are able to signal immediately if any individual animal or a treatment group has lost, or is failing to gain, weight. Such early warnings may be important since weight loss frequently precedes death. Thus weight change should prompt increases in frequency of observations and clinical examinations with perhaps additional blood sampling for haematology and clinical chemistry. Body weight is also necessary for calculating dosage or drug level in the diet.

5.3   FOOD AND WATER CONSUMPTION

Like body weight, food consumption can indicate an adverse effect of a drug at an early stage. Food consumption should be measured before treatment begins. It should also be measured daily in very short term studies, at weekly intervals thereafter. Food consumption data are necessary to calculate the drug : diet ratio if drugs are administered in the diet. A problem with animals which are group-housed is that only group or cage mean food consumptions can be measured.

Water consumption, although not usually routinely measured, should be assessed when conducting studies with diuretics or compounds known, or expected, to affect the kidneys, or if shorter term studies had indicated excessive or reduced drinking. Studies in which dosing is via the drinking water obviously require water consumption measurements.

5.4 OPHTHALMOSCOPY

An ophthalmoscopic examination is used as part of the pretest health check used to select the animals, rodent and non-rodent, to go on test. A similar examination is usually performed at intervals during the study and at the end of dosing. More frequent examinations are necessary if clinical observations suggest evidence of a developing eye defect. In all cases the pupils of the experimental animals are dilated using one or two drops of a sterile mydriatic agent (causing dilation of the pupil). After ten or fifteen minutes the eyes are examined in a darkened room by direct ophthalmoscopy, preferably by a specialist ophthalmic veterinary surgeon or clinician. All results are recorded on ophthalmoscopy report sheets, signed and dated.

5.5 ELECTROCARDIOGRAPHY (ECG)

There are a number of reasons for monitoring ECGs in toxicity studies. These include animal selection procedures to identify animals with cardiac abnormalities and eliminate them from toxicological studies, detection of agents which may produce cardiac effects in the early parts of study without showing later effects, and identification of drugs which may produce ECG changes, e.g. arrhythmias or electrophysiological changes, without causing myocardial lesions.Other reasons include detection of small pathological changes, difficult to find with normal histological techniques, and indication of the time of onset of a cardiac effect.

While ECG measurements can be made in a variety of laboratory animals including the rat, mouse, cat, rabbit and guinea pig, in practice it is the dog which is most widely used, with recordings being made using both limb and chest leads. Measurements are made before the study commences and thereafter at various times during the course of the investigations with all records being examined for abnormalities of heart rate, rhythm and wave form. It is most important to have an experienced person interpreting ECG data since abnormalities, such as isolated ectopic beats, are known to occur in beagle dogs. Other arrhythmias and conduction disturbances which can occur in untreated animals include atrioventricular block, atrioventricular junction beat and, as mentioned above, ventricular and atrial ectopic beat. Thus it is very important to compare pretest results with those recorded during the course of the study.

5.6 BLOOD PRESSURE

In addition to ECG recordings it may be, in certain instances, desirable to record blood pressure. There is now a variety of equipment, available commercially, which will measure and record blood pressure and pulse rates in conscious animals using non-invasive methods. The general principle used for measuring blood pressure in experimental animals is essentially the same as that used in humans, i.e. uniform compression of an artery followed by reduction of the pressure and listening to the blood pressure sounds. However, rather than listening for such sounds with a stethoscope, recordings are now made electronically using a transducer or pneumatic sensor.

To take blood pressure measurements in the rat, the animals are restrained in cages and the measurements made by means of an occlusion cuff which covers the tail. The recording sensors, can be built into the cuff or placed, separate from (under) the cuff, directly on the tail. It is usual to warm the animals by placing them in a thermostatically controlled chamber maintained at 37 °C. This ensures tail dilation, making it much easier to pick up a signal. Examples of equipment which can be used to measure blood pressure in the rat include those supplied by A. H. Horwell Ltd, West Hampstead, London and Harvard Apparatus Ltd, Edenbridge, Kent.

Blood pressure recordings in dogs can also be made using fully automated equipment, often attached to a printer for a permanent record. As with the rat the usual method is to use an occlusion cuff on the tail. In order to ensure the most accurate measurements, it is essential to have uniform compression along at least two thirds of the length of the tail. If the cuff is too small erroneously high blood pressure readings may be recorded.

5.7 LABORATORY PARAMETERS

The analysis of blood and urine samples from animals on toxicity studies is now a common, and indeed required, part of most toxicological investigations (reproductive, *in vivo* genotoxicity and most oncogenicity studies being the notable exceptions). This practice is based on experiences from human clinical medicine, which have demonstrated that changes in blood and urine profiles can provide information on a variety of diseases. Toxicologists now use similar measurements to provide valuable clues and indications of pathological changes occurring as a result of drug treatment. Since, unlike the clinician, we are not involved in making a clinical diagnosis but rather attempting to detect treatment related effects, untreated control groups are almost always included in toxicity studies to provide greater precision in identifying toxicological responses. While 'historic data' from control animals may on occasion be helpful in making such comparisons, because variations in certain parameters can occur as a result of using different strains of animal, age group, nutritional status, etc., it is usually considered essential to include concurrent controls in order to provide the necessary 'normal values'. In addition to individual changes outside the 'normal range', it is also important to look for treatment-related changes within the 'normal limits'. This often assists in the detection of pathological changes at an early stage of their development.

Clinical chemistry and haematological measurements can provide useful information on early toxicological effects, target organ toxicity and reversibility of effects, etc. It is therefore essential that laboratories conducting such measurements use efficient, accurate and reproducible methods, which are not overtly susceptible to interference by extraneous agents such as the presence of the drug and/or its metabolites in blood or urine from treated animals. (This effect must, however, always be considered when attempting to interpret data.) In order to obtain and then retain this precision, most laboratories use automated equipment, e.g. autoanalysers, Coulter Counters, etc. and methods which may have been refined and adapted by that particular laboratory. The inclusion of commercially available 'reference standards' also assists in ensuring the accuracy and reproducibility of results. It is, however, possible that the 'normal ranges' for particular measurements can vary between laboratories, depending on the equipment, method used, etc. Such variation may not be particularly important since comparisons are always made between treated and control data supplied from a single laboratory. It is most unusual for control values from one laboratory to be used as 'limits of normality' for comparison with treatment effects measured in a second.

The sampling techniques used to obtain blood and urine together with the types of measurement performed on these body fluids are discussed below.

5.8 COLLECTION OF BODY FLUIDS
Blood sampling techniques.

In general the collection of blood from dogs and most primates used in toxicology studies causes no particular difficulties. In dogs adequate samples can be obtained from the cephalic, jugular or saphenous veins and in primates the femoral vein is usually chosen. Restraining larger primates may cause some problems and poor technique can be stressful. The difficulty of repeated collection of adequate samples from marmosets is one reason why these species are seldom used for toxicology studies.

For rodents a number of methods can be used. Very large samples can be taken under general anaesthesia from the vena cava, but this procedure is one from which the animals are not permitted to recover. Such collection is used for some pharma-cokinetic and metabolism studies, where groups of animals can be sacrificed at various times. However, in chronic toxicity studies this procedure is only used if, at the end of a study, additional blood is required for measurement of extra haema-tological or blood chemistry parameters.

Cardiac puncture has been used but, although large samples can be obtained, it causes considerable trauma and is not considered an acceptable method for toxicity studies. Orbital sinus bleeds are widely used to obtain large samples (up to 4 ml from rats), but some trauma is caused and it is possible that in studies requiring repeated bleeding, animals may be lost due to eye damage. Also, repeated anaesthesia may exaggerate what, in normal circumstances, may be a slight drug-induced hepatic dysfunction.

Tail tip cutting is often employed to obtain small samples of blood. However, such a method results in samples of inad-equate volume, which are frequently contaminated with tissue fluids. This is especially true for blood samples taken from mice, which are often too small for adequate haematological or blood chemistry measurements. Even in rats something less than 1 ml may be collected, making a complete blood analysis difficult. Also in long term studies rodents can end up with very short tails!

Volumes of up to 2 ml and 0.5 ml can, however, be obtained from the lateral tail vein of rats and mice respectively. In order to obtain such volumes the animal must be first vasodilated, usually by placing them in a warm environment. Adequate vasodilation can be achieved by holding an infra-red lamp directly above the cage. A major drawback with this method is that the animals' environmental temperature cannot be easily controlled and it is possible that they may become overheated. A more satisfactory method is to place the animal cages in a heated, ventilated chamber maintained at about 40 °C. Leaving animals at this tempearture for up to ten minutes is usually long enough to get good vasodilation. A suitable chamber is made by Harvard Apparatus Ltd of Edenbridge, Kent. If only small volumes of blood are required, simply dipping the tail into warm water (37 °C) for one to two minutes will usually cause adequate dilation.

An important point to remember when collecting blood samples is that the sampling site and collection method should all be standardised since both clinical chemistry and haematology values can vary depending on where the blood is obtained (Neptun, Smith & Irons 1985, Eccleston 1977). Thus, for example, the results from a blood sample obtained from the retroorbital plexus should not be compared with those from tail vein blood.

With large animals it is usual to take two blood samples, perhaps a week apart, for haematology and clinical chemistry prior to starting dosing. Such measurements provide a 'baseline' against which further data can be compared. In rodents this may not be possible, or indeed necessary, since most rats and mice used in toxicity studies are inbred, SPF and therefore of a 'standard quality', with only small interanimal variation. Instead of using the test animals, some untreated rodents from the same batch may be sampled to confirm that batch parameters are within the normal accepted range.

In general blood samples are collected at the beginning, during and at the end of treatment, the frequency of such collections depending on the species being used and the duration of the study. Thus, for a 30 day study in rats, blood samples would normally be collected on days 2 and 30, while if the dog were being used an interim day 15 bleed may be included. For studies lasting for 3, 6 or 12, months samples may be collected monthly for the first three months and thereafter probably at three-monthly intervals.

In acute and subchronic studies it is common practice to examine the blood samples to a standard, large battery of laboratory tests. However, once the target organ has been identified, it may be more judicious to include further parameters which will help in monitoring or understanding the toxic response. The sections on haematology and clinical chemistry therefore describe the standard parameters and also some more specialised measurements which may be used to provide further information about toxicological processes.

Urine sampling techniques

As with blood collections, these should be started before the beginning of treatment and repeated at the various intervals during the in-life phase of the study. To collect urine from rats the animals are usually housed individually in 'metabolism' cages for six hours during which time they are allowed access to water, but not to food. Urine samples are obtained from the collection vessels. In some laboratories a 24 hour collection procedure may be used, in which water is withheld during the first six hours. The measurement of the specific gravity of urine passed during the period of water deprivation being considered to provide an index of renal concentration. Collection of adequate samples of urine for analysis from individual mice is only feasible if very small collecting vessels are used, otherwise the urine from several animals can be pooled. Group caged mice can be kept in the same treatment groups in urine collection cages.

Primate urine collection is usually accomplished by attaching a collector to the bottom of the cage. Contamination with faeces and food is, however, frequent and almost impossible to avoid. Dog urine can be collected by placing the animals in specially designed cages, but since dogs frequently avoid urinating for as long as possible in an unfamiliar environment this may result in overnight caging producing no sample. Since samples collected in such cages may also have faecal contamination, a more satisfactory method of collection from dogs is by catheterisation. Urethral catheters made for human use are not very satisfactory for use in beagle dogs, as some damage to the bladder wall can occur. However, infant feeding catheters, 2.5 mm in diameter, cause no trauma and samples of uncontaminated urine can readily be obtained without distressing the dogs.

5.9 HAEMATOLOGICAL MEASUREMENTS

The prime aim of haematology studies (with the possible exception of blood coagulation measurements) is to identify drugs which can exert toxic effects on the cellular constituents of blood, namely red cells (erythrocytes), white cells (leucocytes) and platelets. Such toxic responses can result from the drug having a direct effect on the circulating cells, e.g. haemolytic anaemias caused by red cell destruction, or interference with their production and/or development, e.g. granulocytopenia as a result of drug-induced marrow depression. In either case, the first mode of assessment is to subject blood samples obtained from experimental animals to a variety of basic haematological techniques. For convenience such techniques have been divided into red cell, white cell and platelet studies.

To assess effects on red cells is it usual to perform a red blood cell (RBC) count, haemoglobin (Hb) determination and measure the packed cell volume (PCV or haematocrit). Such parameters provide information on the estimation of the number of circulating red cells, the oxygen-carrying capacity of the blood and the volume of red cells expressed as a fraction of the total blood volume respectively. Using these three methods it is then possible to calculate the so-called 'absolute values', i.e. mean cell volume (MCV), mean cell haemoglobin (MCH) and mean cell haemoglobin concentration (MCHC). Examination of stained blood films can also provide a morphological assessment of the erythrocytes, e.g. are cells large (macrocytic) or small (microcytic), irregular in shape (poikilocytosis), etc., while reticulocyte counts provide an indication of the number of 'juvenile' red cells present in the circulation.

Using data from such studies it is not only possible to detect the development of anaemia, but also possible to gain some insight into the mechanism of such an effect. For example decreases in Hb, RBC and PCV, accompanied by an increased reticulocyte count but normal absolute values, may suggest increased red cell loss or destruction being compensated by normal regeneration. In contrast a decrease in Hb, RBC, PCV and MCHC with a change in red cell morphology (macrocytic) may suggest a macrocytic, iron deficient anaemia.

Routine methods used to assess drug-induced white cell toxicity include measurement of the total number of circulating leucocytes, i.e. white blood cell (WBC) count together with an estimation of the percentage of each of the different types of leucocytes, namely neutrophils, basophils, eosinophils and lymphocytes, present in the sample, i.e. the differential (Diff) count. This last measurement was, until recently, always performed manually, by a skilled operator undertaking a microscopic examination of the blood films, but fully automated differential analysers are now commercially available, e.g. from Hemetrak, Geometric Data, Wayne, USA. By using a combination of these techniques it is possible to determine not only if a drug is reducing (leucopenia) or increasing (leucocytosis) the numbers of circulating white cells, but also if such responses are associated with one or more cell types, e.g. neutrophilia associated with certain inflammatory reactions or eosinophilia caused by parasitic infestation. The separation of neutrophils into different stages of maturity can also be helpful in determining if a leucopenia is due to destruction of peripheral cells or suppression of neutrophil production by the bone marrow. If the blood contains a relatively large number of young immature cells (a so called 'shift to the left') it is probable that the marrow is functioning normally and that the effect is due to destruction in the circulation. Thus once again by careful measurements it is possible not only to detect an effect and its time of onset but also to learn something of its cause.

Platelets (cells derived from megakaryocytes in the bone marrow and essential for normal blood clotting reactions) are subject to a variety of drug-induced changes, including quantitative and functional disorders. Most of these changes involve a reduction in platelet number (thrombocytopenia) which can be detected simply by performing a platelet count. Such drug-induced thrombocytopenia can result from either peripheral destruction or bone marrow suppression. Increased peripheral destruction can result from immune-mediated mechanisms or from platelets having a shortened survival time, while suppression of stem cell levels in the bone causes a reduction in megakaryocytes.

It is possible that a drug may not affect platelet number, but could cause impaired platelet function. In this case platelet counts would be normal, but animals may exhibit abnormal bleeding tendencies, easy bruising, evidence of petechiae (small spots beneath the epidermis due to effusion of blood), etc. However, studies to investigate abnormal platelet function, e.g. platelet aggregation, are not performed routinely.

In addition to examining the cellular components of blood, many laboratories also examine plasma for specific coagulation factors, e.g. prothrombin and partial thromboplastin times. Since there is ample evidence that the liver is the major

organ for the synthesis of most proteins essential for the coagulation system, any increase in prothrombin or partial
thromboplastin times may indicate hepatocellular dysfunction.

By using the pattern of haematological procedures described here it should, in most instances, be possible to detect
gross toxic effects on haemopoiesis (formation of blood) or
mature circulating cells. However, in order to obtain meaningful figures from blood samples it is, as stated previously,
essential to ensure that the laboratory animals on a toxicity
study are standardised in terms of bleeding schedules, housing
conditions, etc.

While we have described some of the more standard studies,
it is possible that non-routine assays would also be performed
to study particular haematological effects. Such non-routine
assays include:

osmotic fragility
plasma viscosity
red cell deformability
methaemoglobin
fibrinogen
fibrinolysis
fibrinogen degradation products
specific coagulation factor assays
platelet aggregation
platelet adhesion
Coombs test (for incomplete antibodies)
haemoglobin electrophoresis

Bone marrow samples, smears and sections, collected at
necropsy are also subjected to a histological examination. It is
possible when conducting toxicological studies in dogs to
obtain biopsy samples of bone marrow under local anaesthesia. Thus any drug effects on precursor cells in the marrow
can also be detected.

5.10 BLOOD CHEMISTRY

Clinical chemistry measurements are of great value to the toxicologist since they can provide early indications of a toxic response, assist in identifying target organs of toxicity and provide information of reversibility of such effects. While some of the measured parameters provide only general information about the well-being of the animal, others may give specific indications with respect to a particular organ. Thus quantitative changes in a particular parameter can reflect

Table 5.1. *Routine clinical chemistry parameters and possible indications of tissue dysfunction*

Parameters	Tissue	General comments
Urea	Kidney	Can be affected by feeding (see Figure 5.1)
Ornithine carbamyl transferase (OCT)	Liver	
Lactic acid dehydrogenase (LDH)	Heart, muscle, liver, RBC	Isoenzymes more useful
Creatine kinase (CK)	Heart, muscle, brain	CK–MB isoenzyme for myocardial injury
Glutamate dehydrogenase (GLDH)	Liver, kidney, brain	Useful for liver dysfunction
Aspartate aminotransferase (AST)	Heart, liver, RBC	–
Alanine aminotransferase (ALT)	Liver, heart	Useful for liver dysfunction
Bilirubin	Liver	Useful in the dog and monkey
Creatinine	Muscle, kidney	–
Albumin	Liver	Albumin : globulin ratio more useful
Alkaline phosphatase (AKP)	Liver, kidney, bone, intestine	Isoenzyme pattern useful for organ specificity

toxicological changes occurring in a particular organ. A list of these 'organ-specific' parameters (usually enzymes) is given in Table 5.1.

Other, less specific, measurements that are routinely carried out include:

    potassium
    sodium
    total protein
    calcium
    glucose

A variety of other measurements can also be undertaken if data for acute and subchronic studies suggest that specific tissues may be affected, e.g. tri-iodothyronine (T_3) and thyroxine (T_4) for thyroid toxicity, and lipase or amylase for pancreatic dysfunction. It is now also possible to investigate general endocrine imbalance in experimental animals by using radioimmunoassays to measure insulin, testosterone, cortisol, adrenocorticotropin hormone (ACTH), prolactin, etc. When performing these measurements it is essential that animals are not stressed, since this can have profound effects on hormone levels. Furthermore, precise timing schedules for blood sampling are essential, since many of these hormones are subject to diurnal variation.

When examining clinical chemistry data, especially enzyme activities, it is important to remember that most of the enzymes are not unique to a specific tissue and also that their relative distributions can vary from species to species (Clampitt & Hart 1978). This does not, however, detract from the usefulness of such measurements in assessing the toxic effect of a drug, especially when it may be possible to differentiate between tissue effects by means of isoenzymes, e.g. differentiating muscle damage from myocardial injury by measuring LDH1 and CK-MB isoenzymes (El Allaf *et al.*, 1986).

In using clinical chemistry parameters to come to some

Table 5.2. *Effect of feeding regime on clinical chemistry parameters in the rat – group mean values*

		ALT (IU/l)	AKP (IU/l)	GLDH (IU/l)	TP (g/l)	ALB (g/l)	AST (IU/l)	CPK (IU/l)	BUN (mmol/l)	CREA (mmol/l)	GLU (mmol/l)	Ca²⁺ (mmol/l)	Chol (mmol/l)	Trig (mmol/l)
A	Mean	81.30	112.65	11.47	71.60	43.95	82.65	200.25	6.85	61.25	7.26	2.60	1.93	2.11
	S.D.	9.73	17.76	4.82	2.64	1.93	9.71	78.49	0.74	5.33	1.15	0.10	0.28	0.71
B	Mean	87.75	109.4	15.38	76.35	44.90	95.45	143.15	8.63	66.65	9.10	2.59	2.20	2.92
	S.D.	19.52	14.60	9.35	2.74	1.74	29.08	45.10	0.78	6.86	1.21	0.11	0.33	0.75
C	Mean	46.63	67.05	12.37	73.56	43.44	91.58	202.53	7.33	61.65	5.87	2.69	2.55	1.36
	S.D.	7.19	8.18	3.17	18.86	2.68	13.0	88.13	1.15	4.50	1.15	0.05	0.35	0.30

Groups of 20 male rats were maintained on different feeding regimes for 2 weeks prior to bleeding.

Group A: restricted feeding 9 a.m. to 3 p.m. (i.e. allowed food only 6 hours per day; starved overnight)

Group B: ad lib fed

Group C: ad lib but starved for 18 hours prior to blood sampling

Results showed no significant differences between groups A and B, i.e. restricted and ad lib fed respectively. Group C, which had not been fed for 18 hours prior to bleeding showed reductions in ALT, AKP, glucose and triglyceride levels.

ALT	alanine aminotransferase	AKP	alkaline phosphatase	GLDH glutamate dehydrogenase
TP	total protein	ALB	albumin	AST aspartate aminotransferase
CPK	creatine kinase	BUN	blood urea nitrogen	CREA creatinine
GLU	glucose	Ca²⁺	calcium	Chol cholesterol
Trig	triglyceride			

conclusion about the toxicological potential of a drug, it is essential to ensure that both treated and control animals have received standardised treatment since differences in measured parameters can result from a variety of reasons other than treatment, e.g. site and time of blood sampling, haemolysis, feeding patterns, etc. Examples of this last effect are shown in Table 5.2 and Figure 5.1. Table 5.2 demonstrates how, in rats, different feeding regimes can influence a number of clinical chemistry parameters and Figure 5.1 shows how, in beagle dogs, blood urea levels are influenced by feeding. Differences in clinical chemistry parameters, between placebo and drug-exposed animals, may therefore not be due to some particular tissue or organ toxicity but may simply be associated with treatment causing depression of appetite or nausea so that dosed animals may not feel like feeding until several hours after the controls.

Fig. 5.1 *Relationship between feeding and blood urea levels in the beagle dog. Blood samples taken from beagle dogs (n = 6), at designated time points i.e. t = 1,3,8 and 24 hours (t = 1 being 9 a.m.), for 3 consecutive days were analysed for urea concentration. Significant increases in blood urea levels were measured 4 hours after feeding (t = 8) returning to base line levels at t = 24. These data show that blood urea levels in the dog can be influenced by feeding patterns.*

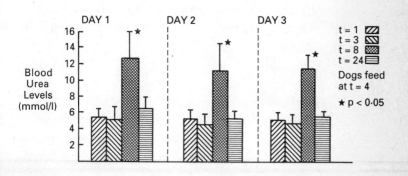

5.11 URINALYSIS

Urine samples obtained from animals on toxicity studies are subjected to a variety of routine examinations which, for convenience, have been divided into physical and chemical measurements. Physical measurements include volume, osmolality, specific gravity, appearance and microscopic examination of sediments. Changes in urine volume, osmolality and specific gravity are usually related to changes in fluid intake, but may reflect increased fluid loss via routes other than the kidney, e.g. emesis (vomiting). In general, reduced fluid intake will result in oliguria, (reduced urination) with urine of high osmolality and specific gravity, whereas high fluid intake will cause polyuria (excessive urination), with low osmolality and specific gravity. Oliguria with low osmolality and specific gravity can, however, occur in acute tubular necrosis where the tubules do not concentrate the glomerular filtrate. Persistent polyuria may also represent impaired renal concentrating ability or pituitary damage.

Urine is usually clear and pale yellow in colour but in certain circumstances it may become turbid and show different coloration. Reasons for turbidity include haematuria, proteinuria, lipiduria, presence of insoluble salts, etc., while abnormally coloured urine could result from the presence of bile pigments (dark green/yellow), haemoglobin (red), coloured drugs, etc. Such changes may not be directly attributable to treatment, e.g. haematuria could be due to trauma caused by the catheterisation process, bitches 'in season', etc., while proteinuria could result from the presence of bacteria or contaminants such as food in the urine.

Deposits obtained following the centrifugation of urine samples are examined for the presence of cells, casts and crystals. The presence of cells, such as erythrocytes and leucocytes (especially polymorphs), can suggest inflammation of the urinary tract. Epithelial cells (in large numbers) could indicate some renal damage, while bacteria and yeasts are often seen in urinary tract infections. The presence of casts and crystals, although often seen in urinary deposits from healthy untreated animals, can indicate some renal effects.

Chemical assessment of urine usually involves measuring protein, reducing sugars such as glucose, electrolyte (Na^+ and K^+), enzymes (e.g. N-acetyl glucosamine and alkaline phosphatase) and urinary pH. Increased proteinurea is almost invariably associated with renal damage with the degree and type of response possibly providing some clues to the anatomical site of the damage, e.g. glomerular dysfunction is associated with high molecular weight proteins and tubular damage with low molecular weight proteins. Measurement of glucose and electrolytes can also provide some information on kidney damage. However, it must be remembered that electrolyte concentration can be strongly influenced by salt and fluid intake. Enzymuria can also indicate functional abnormalities, but increased alkaline phosphatase activity may be associated with prerenal damage, e.g. hepatotoxicity.

Urinary pH is invariably measured, and while dog and rat urine is usually acid, wide ranges of pH can be measured, i.e. 4.8 to 8.5 in normal untreated animals. Moderate changes can indicate acid–base and K^+ status, but presence of certain drugs or bacteria can have a similar effect. Less common assays include those for ketones and urea, while bilirubin, bile salts and urobilinogen may have some use in assessing liver function.

Urinalysis can provide a useful and simple procedure for detecting renal damage. An important feature of such measurements, however, is that urine samples should be collected very early in a study preferably following only one or two days' treatment. The reason for this is that the kidney has a remarkable compensatory capacity; thus, while the first treatment can produce profound changes in urinary parameters, the organ can soon compensate for the damage and, even with continued treatment, subsequent urine samples may

Table 5.3. *Urinary changes (mean ± SEM) in male rats[a]*
treated with mercuric chloride (2 mg/kg/day) for 10
consecutive days (urine collected on days 2 and 8)

Urinary parameter

Group	Total protein (mg/l)	Glucose (mmol/l)	Total AKP	Volume (ml)
Day 2				
1	368	0.44	1691	2.92
	±88	±0.36	±741	±0.69
2	826*[b]	19.3 *	5939*	3.37
	±285	±9.9	±2356	±1.02
Day 8				
1	408	0.29	1136	2.16
	±115	±0.10	±516	±0.76
2	632	1.78	1031	3.17
	±337	±2.10	±548	±0.61

Group　Treatment ($N = 20$)
1　　　　10 ml distilled water per day
2　　　　Mercuric chloride (2 mg/kg)
Urine samples were collected over a six-hour period, collection
beginning immediately after dosing.
[a] Female rats showed similar response
　　(Results courtesy of Mr G Salmon, Department of Toxicology,
　　Smith Kline & French)
[b] $p < 0.05$ significance level

show no abnormal measurements. An example of such an
effect can be seen in a study undertaken with the nephrotoxic
agent mercuric chloride. Rats were dosed (2 mg/kg body
weight/day) for ten consecutive days with the chemical and
urine samples collected on days two and eight. While large
changes in certain urinary parameters were seen on day two,
by day eight these had almost all returned to normal (Table
5.3.). At necropsy, on day 11, the kidneys from the mercuric
chloride animals were found to be significantly larger than
those from the controls, i.e. 1.27 ± 0.2 g as compared to 1.0
±0.1 g, and showed histopathological changes to the proximal

tubules. This short study amply demonstrates the importance of collecting urine samples early in a study, otherwise effects may be missed due to compensatory mechanisms.

5.12 DRUG ANALYSIS

It is now becoming a routine part of certain toxicological investigations, especially 30 day studies, to take blood samples for measuring blood drug concentrations. (Such analyses may be performed on whole blood, plasma or serum and may include measurement of parent compound and/or metabolites.) The usual procedure is to produce a fairly comprehensive blood drug profile, perhaps ten or twelve sampling times, on the first and last day of treatment. Such measurements provide an opportunity to relate toxic responses to blood drug concentrations and compare pharmacokinetics in 'naïve' animals (first day of dosing) and after 30 days of dosing, e.g. detection of enzyme induction, drug accumulation, etc., comparison of metabolic and pharmacokinetic profiles with dose, e.g. do blood levels increase correspondingly with dose, are different metabolites formed at high doses, etc?

It is of course possible that if 'large' blood volumes are collected frequently, then the toxicity study may become 'compromised', e.g. animals may become stressed, especially in high dose groups. In practice, however, this should not occur since it is possible to take repeated blood samples from rats and dogs, 0.1 and 4.0 ml, respectively, for drug metabolism and pharmacokinetic (DMPK) analysis, without adversely affecting clinical chemistry or haematology parameters. Other species, such as monkeys, can become stressed with repeated bleeding and certain haematology parameters, especially red cell counts and haemoglobin, can drop rapidly, often requiring several days or even weeks to return to normal. In general terms, however, by using careful sampling procedures, it is possible to obtain data on blood drug concentrations from both rodent and non-rodent species, and perhaps determine if the toxicological susceptibility of a particular species is related to such measurements.

In conclusion these lists of stereotyped measurements which are made, together with the equipment and techniques used, should not mislead anyone into believing that conducting toxicity studies simply involves following a 'recipe'. Much thought should be given before each study starts to ensure that what is proposed is relevant to that particular drug. Consideration must always be given to background data which are available from shorter term investigations, studies in other species or studies with related compounds, which may result in the incorporation of additional or alternative assays. The limited volumes of body fluids which can be humanely obtained from small laboratory animals make it necessary to choose the priorities of each investigation very carefully.

REFERENCES

Clampitt, R. B. & Hart, R. J. (1978) The tissue activities of some diognostic enzymes in ten mammalian species. *Journal of Comparative Pathology*, **88**, 607–21.

Eccleston, E. (1977) Normal hematological values in rats, mice and marmosets. In *Comparative Clinical Haematology*, ed. R. K. Arches & L. B. Jeffcott. Blackwell, Oxford, pp. 611–19.

El Allaf, M., Chapelle, J. P., El Allaf, D., Adam, A., Faymanville, M. E., Laurent, P., & Heusghem, C. (1986) Differentiating muscle damage from myocardial injury by means of the serum creatinine kinase (CK) isoenzyme MB mass measurement/total CK activity ratio. *Clinical Chemistry*, **32**, 291–5.

Irwin, S. (1962) Drug screening and evaluative procedures. *Science*, **136**, 123–6.

Neptun, D. A., Smith, C. N. & Irons, R. D. (1985) Effect of sampling site and collection method on variations in baseline clinical pathology parameters in Fischer-344 Rats. *Fundamental and Applied Toxicology*, **5**, 1180–5.

6

Terminal studies

6.1 NECROPSY

At the end of treatment, usually 24 hours after the last dose, the experimental animals (controls as well as those treated with the test material) are given a final thorough examination for any external abnormalities and then killed. The method used to kill the animals depends upon the species on test. Rats and mice are usually deeply anaesthetised by exposure to CO_2 gas or an overdose of diethylether vapour, while larger animals such as dogs and monkeys may be sedated with acetyl promazine before being anaesthetised by an overdose of intravenously administered sodium pentabarbitone solution. When it is certain that the animals are deeply anaesthetised and insensitive to pain, i.e. cessation of palpebral reflex (in normal conscious animals if the area around the upper eyelid is gently touched the animal will blink; however, in deeply anaesthetised animals this reflex disappears), they are usually exsanguinated. In rats, mice and in some monkeys this can be accomplished by using a syringe to withdraw the blood, usually via the posterior vena cava, whereas dogs are usually exsanguinated through the axillary blood vessels.

Once the animal is dead it is subjected to a *port-mortem* examination when the organs and body cavities are given a thorough macroscopic examination for signs of drug-induced and/or spontaneous lesions. (When conducting a necropsy it is sensible to have a summary of in-life observations available, since such records can often prove useful in identifying lesions.) At present, in most laboratories macroscopic abnormalities are recorded using descriptive terms, e.g. shape, size (measurements when appropriate), texture, colour, etc.,

Table 6.1. *List of tissues normally collected at necropsy*

Adrenal glands	Muscle (skeletal)
Aorta	Nerve (sciatic)
Bladder, urinary tract	Oesophagus
Bone	Pancreas
Bone marrow	Optic nerve
Bone marrow smear	Ovary
Brain	Pituitary gland
Caecum	Prostate gland
Cervix	Salivary gland (mandibular)
Colon	Salivary gland (parotid)
Duodenum	Salivary gland (sublingual)
Epididymis	Skin
Eye	Seminal vesicles
Gall bladder	Spinal cord (cervical)
Heart	Spinal cord (lumbar)
Ileum	Spleen
(Injection site)	Stomach
Jejunum	Testes
Kidney	Thymus
Liver	Thyroid gland
Lungs	Tongue
Lymph node (mandibular)	Trachea
Lymph node (mesenteric)	Uterus
Mammary gland	Vagina

All tissues are fixed in neutral buffered formalin, except for eyes, bone marrow smear and testes/epididymides, which are fixed in Davidson's solution, methanol and Bouin's solution respectively. The eyes are fixed in Davidson's solution for 24 hours before being transferred to 70% alcohol. The testes/epididymides are kept in Bouin's fixative for 48–72 hours before also being transferred into 70% alcohol.

directly onto special colour coded, pathology record forms, which are then retained as a permanent record of the post-mortem observations. Examples of unusual findings may also be photographed and included in the raw data. In the future, however, necropsy data will usually be recorded directly onto a computer. In some laboratories such a procedure is already standard practice.

During the necropsy procedure a standard selection of organs and tissues are removed from the body cavity and stored in a preserving fluid (fixative) for later histological examination. (A standard list of such tissues is shown in Table 6.1.) Even if it is not the intention to process all the tissues for histological examination, it is often considered prudent to take all the tissues in fixative. Such preserved material will remain intact for many years and can therefore be examined at a later time if any problems arise. Thus the simple act of removing and storing tissues may save the necessity of performing a repeat study. A problem with this approach is that keeping possibly unneeded preserved tissues takes up storage space and requires comprehensive recording systems to satisfy GLP requirements. Since it is undesirable to squander resources by taking and keeping tissues surplus to requirements, the decision on which tissues to take at necropsy should be made on an individual case basis.

It has been argued that the use of standard comprehensive lists is unnecessary and wasteful since the majority of tissues are usually found to be normal. Various suggestions have been put forward, for example that only the 'major organs' (e.g. heart, kidney, liver) and macroscopically abnormal tissue should be taken at necropsy, thereby saving effort on microscopic examination. Such restricted examinations are, however, questionable since it is possible that tissues may have cellular changes not evident from a macroscopic examination. Also, it is not possible to predict which are the 'major organs' in terms of a toxicological response; thus, while the liver, kidney, etc. may be histologically normal, other tissues such as thyroid may show histopathological changes. In general, comprehensive lists of tissues should be collected and examined in an attempt to 'cover all aspects' of tissue toxicological response. As well as these 'standard tissues', any other abnormalities and macroscopic lesions should also be taken for subsequent microscopic examination.

In many instances it is not possible to take an organ in its entirety, simply because of its size and the difficulties encountered in getting adequate fixation quickly enough. In such cases, to ensure samples of tissues are treated in a consistent manner, it is usual to have a trimming procedure. The anatomical location from which the tissue section will be removed is described, e.g. for liver sections from the left lateral, caudate and median lobes may be stipulated; kidney – transverse section through the central portion of the renal pelvis. In all cases the desire is for uniformity of tissues in order to make comparison between animals more effective.

Animals which die unexpectedly, or are in a moribund condition which necessitates them being killed prematurely, also receive a complete post-mortem examination with all tissues being collected. If animals are, however, severely autolysed (their tissues self-digested), they are usually necropsied and inspected for obvious changes but the tissues are not normally collected.

To perform necropsies accurately it is necessary to have individuals of the highest level of expertise and competence who are expert in detecting changes in tissue. Accordingly, all necropsy personnel must have received a thorough training, and during all necropsy procedures it is essential to have a senior scientist present who is capable of 'interpreting' suspected abnormalities and macroscopic lesions, which should be brought to his/her attention.

The order in which tissues are taken is not fixed by regulations and indeed slight variations may exist in the same laboratory depending on the individual performing the necropsy. In general, however, it is considered advisable to remove as quickly as possible those organs which undergo rapid autolysis, e.g. eyes, gut, etc. Naturally, when samples are required for ultrastructural or histochemical analysis, it is important to take these tissues first.

In addition to recording macroscopic changes, the measurement of organ weights is now a common part of most necropsy procedures. The pros and cons of making such measurements are still, however, often discussed, as there are cogent arguments both for and against weighing organs. Those against organ weight measurements argue that collecting such data is a time consuming and wasteful procedure, in terms of data handling, checking tables for mistakes, etc., since it provides only limited if any, useful information. In fact it can be argued that such measurements may even give rise to false results if the tissue is inadequately trimmed (removal of surrounding connective tissue) or if pieces of the tissue to be weighed are accidentally cut off during trimming. In practice, however, although such problems can arise, especially when handling small tissues, e.g. rat thyroid glands, they can be overcome by using only well-trained staff for such procedures. Presentation of organ weight data can also prove something of a problem. Just showing relative organ weights can be misleading if treatment is associated with reduced body weight gains but leaves the organs comparatively unaffected. Thus, in order to come to any decisions, it is necessary to have not only organ weights, but also body weights and perhaps food consumption data. It has been suggested that more standard and reproducible measurements other than body weight should be used to calculate relative organ weights. Such standard measurements in the rat include skeletal measurements, e.g. femur length, and brain weight.

Some tissues can be affected simply as a result of the animal being stressed, e.g. the thymus. Thus changes in thymus weight may reflect a non-specific, stress-related phenomenon rather than a direct target tissue effect. The maturation of sex organs can vary between similarly aged animals, especially dogs, thus an apparent reduction in testes or prostate weight may not be related to treatment, but simply reflect the fact that the animals in a treatment group are perhaps not as sexually mature as those in the control group. Although this effect should be overcome by randomisation or using sexually mature animals, such problems are not uncommon. These problems of interpretation, it is argued, are easily overcome if changes in organ weights are accompanied by histopathological effects. However, if histopathological responses are present then changes in organ weights become of secondary importance.

It is possible that changes in organ weights may reflect tissue changes not readily detectable by microscopy, e.g. a 10% increase in liver weight is readily measurable but a 10% increase in the size of a hepatocyte is not so easily recognisable. Thus, in mild hypertrophy, measurements of organ weights can supply useful information concerning treatment-related effects. It is therefore argued that although routine organ weights when considered in isolation may not contribute significantly to the interpretation of toxic effects, they can, when correlated with clinical or biochemical changes, provide the pathologist with early information to help in identifying target tissues. Since organ weights are inexpensive to record and easy to perform they are now a standard inclusion in toxicity studies. However, the fact that a measurement is easy to make does not justify its inclusion in a scientific investigation. In our opinion, if there is a good scientific reason for recording the weight of a particular organ then this should be performed, but simply taking measurements because they are considered cheap and easy is not good science.

Regulatory attitudes towards recording weights vary quite considerably. While the EEC do not specifically mention recording organ weights, other guidelines, e.g. Australia, Canada and OECD, ask for this information without specifying which measurements they think should be made. The Japanese, however, are quite precise, and many consider unreasonable, in their demands requiring that the liver, kidneys, heart, brain, thyroids, testes, ovaries, prostate, spleen and pituitary, the submaxillary salivary gland, thymus, lung, epididymis and uterus are weighed. There is, however, confusion as to whether such measurements are really required in the absence of any histological or biochemical evidence of a toxic effect.

6.2 HISTOLOGY

Tissues taken at necropsy are usually processed into stained paraffin wax sections for subsequent microscopic examination by an experienced pathologist. The first step in such histological proceedings is to ensure that the tissue taken at post mortem is placed in an adequate volume of an appropriate fixative as soon as possible after the death of the animal. Fixation stabilises the tissue for subsequent treatment, while preventing autolysis and putrefaction. To ensure good preservation of cellular detail, ideal fixatives should act rapidly and uniformly throughout the tissue and not cause shrinkage or swelling. It is usual that the amount of fixative exceeds the mass of tissue samples by a ratio of at least 5 : 1, and that tissues are allowed to stand for about three days before being prepared for histology.

When the tissue has been thoroughly fixed, it is usually embedded in paraffin wax. This is achieved by dehydrating the tissue in ethanol, which is then removed in a clearing agent such as chloroform, which in turn is removed in a bath of molten paraffin wax (an example of a routine processing schedule is shown in Figure 6.1). This process is now usually automated, using one of a range of commercially available processing equipment. When the tissue has become thoroughly impregnated with wax it is left to cool and solidify.

Fig. 6.1 *Example of a routine tissue processing schedule.*
(Tissues containing calcium salts, e.g. bone, teeth, etc., are too
hard to be sectioned by conventional techniques and are there-
fore decalcified after fixation but before dehydration. Decalci-
fying fluids which remove calcium salts by dissolution include
Custer's fluid (20% aqueous trisodium citrate and 90% formic
acid in a ratio of 1 : 1 by volume), 5% aqueous trichloroacetic
acid and nitric acid formalin.)

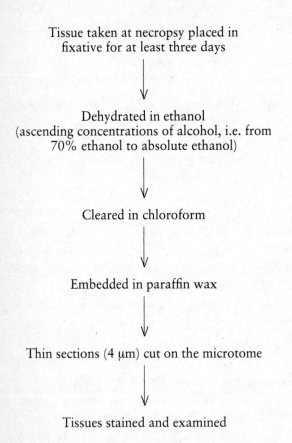

Tissue taken at necropsy placed in
fixative for at least three days

Dehydrated in ethanol
(ascending concentrations of alcohol, i.e. from
70% ethanol to absolute ethanol)

Cleared in chloroform

Embedded in paraffin wax

Thin sections (4 μm) cut on the microtome

Tissues stained and examined

On cooling, the surplus wax is removed from the face of the block to reveal the tissue. Thin sections of the tissue, nominally 4 μm, can then be cut on a microtome (a precision instrument especially designed for cutting materials into sections) for subsequent staining and microscopic examination.

In order that the pathologist can determine cellular detail, it is necessary to stain the tissue sections. However, since wax prevents penetration of dyes it must first be removed in a solvent, usually xylene. For routine morphological assessment the dyes haematoxylin and eosin, which stain the nuclei and cytoplasm respectively, are most generally used. More specific histological techniques are also available to demonstrate various tissue components: specific proteins (e.g. collagen), pigments (e.g. bile pigment), carbohydrates (e.g. glycogen), etc. In the majority of cases, once the section has been stained a coverslip is then attached over its surface with a viscous fluid called a mountant. The coverslip protects the tissue and allows a better microscopic examination.

There are various other specialised histological processes, including enzyme histochemistry, immunocytochemistry and electron microscopy, available to the pathologist. Since such techniques are now playing a greater role in toxicological studies they are worthy of a brief description.

When using techniques such as electron microscopy or enzyme histochemistry it is essential to ensure that once the animal has been killed the tissue to be examined is preserved as rapidly as possible. Such rapid preservation helps in preventing any loss of structural architecture or enzyme activities, which occur rapidly after death.

For electron microscopic examination the tissue is usually fixed by perfusion using appropriate equipment (see Fig. 6.2). Electron microscopy is used to provide better definition of morphological problems that have already been identified by light microscopic examination. Such information, possibly on specific organelle effects, can provide valuable clues to the

Fig. 6.2 *Example of a routine tissue processing schedule for electron microscopy.*

Tissue perfused via major blood vessel with isotonic saline at a constant pressure of 25 mm Hg (this removes the majority of blood allowing better fixation)

Perfused with fixative e.g. 4% glutaraldehyde in 0.1 M cocodylate buffer (pH 7.4), again at a pressure of 25 mm Hg for about 10 minutes.

Cut into small pieces (1–2 mm diameter) and immersed in 4% glutaraldehyde in 0.1 M cocodylate buffer for a further 1–2 hours

Washed in buffer and post-fixed in 1% osmium tetroxide solution

Dehydrated in ascending ethanol series and 1,2-epoxypropane before being embedded in resin

Semithin sections cut (nominally 1 μm) to select areas for ultrastructural examination

Ultrathin section cut (70–90 nm), stained with uranyl acetate and lead citrate before examination in an electron microscope

mechanism of toxic effects. The major drawback of electron microscopy is the limited numbers of samples which can be processed and examined, although correct use of semithin sections can help in identifying areas of interest.

For enzyme histochemistry it is essential that the tissues collected are as fresh as possible and frozen as rapidly as possible. The usual procedure is to place the tissue of interest into dichlorodifluoromethane (precooled in liquid nitrogen). The frozen tissue can then be stored in liquid nitrogen or even freeze-dried. For enzyme histochemical studies, the frozen tissues are cut with a special microtome called a cryostat and the still frozen section is processed to demonstrate the presence of enzyme activities. The section can then be examined manually or automatically using a scanning microdensitometer which gives an integrated density reading of the cells being examined, thus providing a quantitative estimate of enzyme activity.

6.3 HISTOPATHOLOGICAL ASSESSMENT

It is essential that the microscopic examination of the tissues from treated and control animals be conducted by a qualified, experienced and competent pathologist. The need for an accurate assessment is obvious when it is remembered that such a judgement may decide whether or not to proceed to clinical trials or marketing. Thus the failure to detect even subtle pathological effects may inadvertently expose human volunteers or patients to a potential risk. On the other hand, it is important that valuable time and resources are not squandered on irrelevant tissue changes which have no pathological implications.

In order for the pathologist to come to an accurate diagnosis it is essential to have all relevant data, including clinical observations, macroscopic necropsy findings and haematology and clinical chemistry measurements, available. It has been argued that such information, together with knowledge of animal treatment, may introduce some sort of bias in assessment and it has been suggested that 'blind slide' reading should be introduced. However, to come to a balanced assessment it is necessary to have all the available data to hand. If any problems of judgement or interpretation do occur it is always possible to review the appropriate slides using a properly designed blind trial.

The minimum routine histopathological assessment of rodent toxicity studies must ensure that the pathologist examines all tissues from the high dose and control groups. When lesions are seen, the same tissues from the middle dose group are then examined and, if the effect is still present, the low dose slides will also be examined. Such an approach identifies organs affected by treatment and the dose at which the effects are no longer seen. An exception to this procedure is the case of oncogenicity studies where it is common practice to examine all tissues from all animals in all groups.

When using dogs, monkeys, etc., it is usual that all tissues from all animals are examined. The reason for this is that only low numbers of dogs or monkeys are usually used. However, if no effects are seen in a tissue from a high dose animal there would appear to be little point in examining the same tissue from animals treated with lower doses.

6.4 DATA ANALYSIS

It may be pertinent in the chapter discussing terminal studies to devote some time to the ultimate part of any toxicological study which is the description, presentation and discussion of the results of the investigation in a final report. To many of us, producing reports is perhaps the most unsatisfactory and least enjoyable part of a toxicological investigation, and as a result is often done badly. It must be remembered, however, that science is about communication and the most thorough and interesting work is meaningless if no-one can understand it or, because of the way it is presented, it is never read. As toxicologists we do not manufacture items such as motor cars or televisions, therefore the only tangible evidence of our work is the written report and it is upon this that we will be judged. In order to produce clear, succinct reports it is necessary to carry out a critical review of each study. This must bring together all pertinent facts which give a guide to the toxic potential of the test agent, while avoiding confusion with isolated events which have no relevance to the intrinsic toxicity of the chemical.

As we have seen in the preceding chapters, at the end of a study it is probable that a large amount of data will have been amassed which must be presented in a logical form. There is a tendency therefore to become obsessed by statistical techniques and require that all data be subjected to 'statistical analysis'. Statistics are, however, tools to answer particular questions with and not to go 'fishing' with. Lord Rutherford was perhaps being a little harsh when he stated: 'If your experiment needs statistics, you ought to have done a better experiment.' On the other hand, biological systems are extremely complex and using statistical analyses can often help in understanding and interpreting data. The use of extensively complicated statistics is, however, a poor substitute for good study design and clarity of thought and toxicologists should, in many instances, be able to interpret the data without the assistance of statistics.

It is not our intention to describe the various statistical methods, or how and when they should be used, since this topic has been the subject of a number of books (e.g. Armitage 1971, Batschelete 1979, Peto *et al.* 1980, Bailey 1981, Salsburg 1981). It is perhaps worth suggesting to the reader that visual presentations of analyses, such as plotting results in the form of suitable graphs, histograms, scatter plots, etc., can often enhance clarity and aid understanding.

REFERENCES

Armitage, P. (1971) *Statistical Methods in Medical Research*. Blackwell, Oxford.

Bailey, N. T. J. (1981) *Statistical Methods in Biology*, 2nd edition. Hodder & Stoughton, London.

Batschelete, E. (1979) *Introduction to Mathematics for Life Scientists*, 3rd edition. Springer-Verlag, New York.

Peto, R., Pike, M. C., Day, N. E., Gray, R. G., Lee, P. N., Parish, S., Peto, J., Richards, S. & Wahrendorf, J. (1980) Guidelines for simple sensitive significance tests for carcinogenic effects in long term animal experiments. In *Long-term and Short-term Screening Tests for Carcinogens: A Critical Approach*. IARC Mongraphs on the Evaluation of Carcinogenic Risk to Humans, Suppl 2, 311–426.

Salsburg, D. S. (1981) Statistics and toxicology: an overview. In *Scientific Considerations in Monitoring and Evaluating Toxicology Research*, ed. E. J. Gralla, 123–36.

7

Specialised routes of exposure

Regulatory authorities generally require, as a valid predictive toxicological investigation of a new medicine, that toxicity studies are conducted using the route of administration which is to be used clinically. This requirement can, from time to time, prove problematic for the toxicologist. For instance, while it is usually possible to dose compounds by the proposed route of clinical administration, it is often impossible, both in terms of quantity and duration, to treat animals with large multiples of the human dosage. For example, there are physical and humane limitations on the maximum volumes that can be administered intravenously, intramuscularly, subcutaneously, intraperitoneally or into the rectum, vagina, eyes, etc. Increasing drug concentrations to multiples of the human dose may be similarly restricted because of problems with solubility pH, tonicity, etc. For some routes of administration it may therefore be necessary to achieve the overall objective of predicting human risk by somewhat indirect means. For example, to assess the systemic toxicity of a topically applied compound, it may be possible to measure the maximum drug blood levels achieved after local application and then, for longer term studies, use a second route of administration with doses giving equivalent, or even higher, systemic exposure than that achieved by the local route. Thus, for a compound which might be given clinically intramuscularly, intravenously, intraperitoneally or subcutaneously it would not be technically possible, or indeed useful, to conduct long term toxicity studies by these routes. If, however, it was found to be well absorbed following oral administration, then oncogenicity studies could be conducted using this route provided, of course, that adequate drug blood levels were achieved.

In this chapter some specialised routes of administration are considered individually. The description of these methods will, however, be truncated since the inhalation route, for example, has developed such an extensive technology that a whole book could be devoted to this method of administration alone.

7.1 INTRAVENOUS

It is quite common to conduct toxicological studies using IV administration, even though this may not be the intended clinical route. There are a number of reasons for this. First, it gives a 'measure' of the toxicity of the drug uninfluenced by such factors as absorption or first passage through the liver. This is particularly important in acute toxicity studies where a comparison of the acute toxicity by the oral and IV routes may give an indication of whether or not there are problems with destruction of the drug in the gut, poor absorption, etc. Second, many clinicians investigating human metabolism, pharmacokinetics and pharmacology prefer to use an IV infusion rather than the oral route. This is because an infusion can be stopped immediately at the first signs of any untoward effect, whereas if a drug has been taken by mouth further absorption can still occur *after* the first indication of adverse effects.

A few drugs are, however, developed specifically for clinical use by the IV route. Since these are usually designed to be used in patients over relatively short periods of time, most requirements for IV toxicity studies only ask for treatments to last a maximum of three months. Such studies may use bolus injections, in which the drug is administered rapidly over a period of a minute or two, or infusions, when it is given more slowly, often over many hours. The latter method of administration presents more technical problems than the former, since it becomes necessary to restrain the animal while the infusion is proceeding. Also, if the infusion is over a prolonged period of time, e.g. 24 hours, it is possible that blood clots may begin to form at the tip of the cannula. Such thrombus formation can reduce blood flow, causing the drug to seep into the interstitial spaces which may, when using irritant compounds, produce severe inflammatory reactions. Since different problems can arise depending on the species being used, a brief description of the use of dogs, primates and rodents in IV studies is given below.

In general IV dosing studies in dogs should present no particular difficulties, although it is important to select tractable animals and to have a quiet area away from the animal pens where the injections may be given. The usual procedure for giving bolus injections to dogs is to have one person restraining the animal while a second administers the injection. It is preferable to have the holder sitting next to the dog on a low bench, so that the individual giving the injection can do so while standing facing the dog (Figure 7.1). The cephalic veins are usually used on alternate occasions but the saphenous veins can also be used, particularly in long term studies. The usual method is to shave the leg over the vein, swab the area with 70% ethanol or surgical spirit and give the injection using a fresh sterile disposable needle (19 G, 38 mm in length being most appropriate).

For infusion studies it is usually best to restrain the dog in a standing position in a 'Pavlov' type sling attached to a purpose-designed stand such as the one illustrated in Figure 7.2. These frames have the advantage that there are no corner uprights to restrict free access to the legs of the dog. Apparatus, such as infusion pumps, can be placed on a lower level platform out of reach of the dog, which is held by a close fitting zipped jacket clipped onto the horizontal beam. Sterile intravenous catheter sets such as the 'Portex' 1/D 0.5 mm, 18 G are suitable for most infusion studies in dogs, while variable speed infusion pumps able to utilise a range of syringe sizes (such as those manufactured by IMED, Abingdon, Oxfordshire) allow a range of infusion rates and dose volumes to be given. Before starting an infusion it is essential to ensure that the catheter is full of solution thus preventing the risk of an air

Fig. 7.1 *Technique for giving a bolus injection to a dog.*

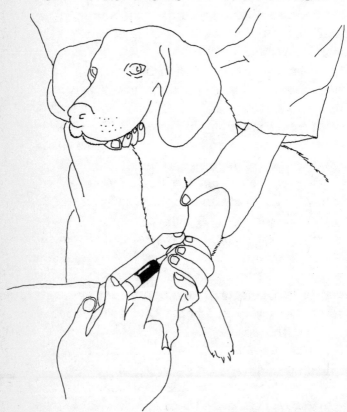

embolism. Using such techniques, IV infusions of up to an hour in duration can be given daily for periods lasting up to 12 months, while single infusions lasting 24 hours cause no difficulties. Where longer continuous infusions are required, it is usually necessary to implant a catheter surgically and use ambulatory infusion units.

When undertaking IV studies in primates it is usual to use larger animals, e.g. cynomolgus, rhesus monkeys or baboons, smaller monkeys such as marmosets are not suitable. Because monkey veins are easily damaged, it is usual to restrict periods of treatment to three months' duration or less. For bolus injections the principles are the same as for the dog, with one operative restraining the animal while a second gives the injection. Several sites can be used, but the saphenous vein is perhaps the most popular. For infusion studies, monkeys are usually restrained in specially designed chairs, such as those supplied by Plastics Manufacturing and Supplies Inc., Lansing, Michigan, USA. However, because of the stress caused by such restraint, IV infusions in monkeys should be limited to a maximum duration of one hour daily. It is possible to conduct

Fig. 7.2 *Use of 'Pavlov' type sling for infusion studies in the dog.*

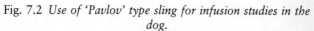
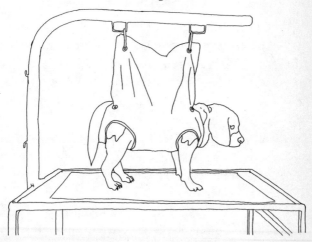

single IV infusion studies for longer periods of time and 24 hour continuous infusions are conventionally undertaken with cynomolgus monkeys.

Intravenous dosing in rodents presents some technical difficulties because of the small size of their veins. Injections in rats and mice are best made into a lateral tail vein using a 25 G, ⁵/₈ inch needle. The vein should have been previously dilated either by placing the animals in a warm environment as described in Chapter 5, or simply by dipping the tail in water at 37°C for a minute or so. It is also good practice to use different veins on alternate days. For mice, studies of 30 days' duration are possible, while for rats much longer studies can be undertaken although studies of more than 3 months are uncommon. Infusion studies can also be undertaken in both rats and mice; dosing being in restrained animals using sterile disposable butterfly infusion sets. (The butterfly and the tail are immobilised during infusion by taping them to the bench.) Infusions lasting less than an hour may be given daily and single infusions of 12 hours are feasible. For longer periods of infusion, surgically implanted in-dwelling catheters may be preferable.

7.2 INTRAVENOUS/PERIVENOUS

Prior to the first IV treatment of human volunteers, it may be necessary to determine any possible local irritancy of the clinical formulation. (This is a requirement of the Federal German Health Office.) Studies are therefore often undertaken to determine whether leakage or inaccurate dosing of the drug (e.g. if the needle goes through the vein) will cause perivenous irritancy. The usual procedure is to take two beagle dogs and allocate each leg to a particular treatment. The cephalic and saphenous veins are then injected both intravenously and perivenously with the drug or a placebo solution (matched in pH, tonicity, etc. to the test material). An example of such a dosing arrangement is shown in Table 7.1. Approximately 48 hours

Table 7.1 *Comparison of intravenous and perivenous injections in Beagle dogs*

Dog number	Treatment			
	Front legs (cephalic vein)		Hind legs (saphenous vein)	
	Left	Right	Left	Right
1	D	C	C	D
2	C	D	D	C

C = control solution, D = drug solution

after dosing, the dogs are killed and the injection sites examined macroscopically before being removed and placed in neutral buffered formalin for histological processing.

7.3 INHALATION

Dosing by inhalation, while presenting certain technical problems, does provide a number of advantages over the more conventional routes of drug administration, e.g. rapid onset of drug action, low incidence of side effects, smaller amounts of drug required for dosing, etc. Since drugs to be given by inhalation are, with the exception of gaseous or volatile anaesthetics, usually formulated into aerosols, this brief description of inhalation toxicology includes some simple concepts of aerosol terminology and describes some of the technical considerations which must be borne in mind when administering drugs by inhalation. In general terms an aerosol can be described as a two-phase system consisting of gas and particles, with the particles being either liquid (as in a mist) or solid (as in a dust). An aerosol should not be confused with a vapour which is a single gaseous phase. Vapours can, however, be transformed into aerosols by condensation and, similarly, aerosols into vapours by evaporation or sublimation. Thus an aerosol formulation can be affected by temperature, humidity, flow rate, etc. A consideration with all aerosol formulations, irrespective of whether the particulate material is liquid, solid or a combination of the two, is the actual size of the particles. This is because it is the size of the particle which ultimately determines where in the respiratory tract it will be deposited. For nasal respiration it is considered that particles greater than 5.5 µm in diameter are non-respirable, particles between 5 and 1 µm settle out in the small airways, while particles less than 1 µm in diameter reach the alveoli. Thus the smaller the particulate size, the deeper the penetration into the lung. Particulate retention in the upper respiratory tract does, however, vary widely, between species, being greatest in small rodents such as mice and rats.

Generation of stable homogeneous particles requires the use of specialist apparatus, such as atomisers or aerosol cans for producing mists, while dust clouds can be generated using a variety of techniques. For example, the Wright generator utilises a rotating blade which scrapes impacted fine powder from the surface of a packed cylinder which is then made into an aerosol by compressed air. Fluid bed aerosol generators, such as that described by Willeke, Lo & Whitby (1974), use a fluidised bed of beads which, by means of impaction and turbulence, produces a stable output of particles. Some methods also incorporate elutriation devices (which separate coarser particles from finer ones), and with some powders it may be necessary to remove any electrical charge before the aerosol is finally released. It is essential to monitor the generated aerosol (as near to the animal's nose as possible) to ensure that particle aggregation has not occurred and that the aerosol remains stable and homogeneous. (This involves another area of technology which is too broad-ranging for consideration here.)

Naturally in order to achieve the desired therapeutic effect it is necessary not only to get an adequate amount of drug into the lungs but also to ensure it reaches the desired location. Variations in the surface concentrations of drug at different sites in the lung are determined by the anatomy of the lung and particle size. Drug concentrations are highest in the central airways and, as a result of the large increase in surface area, decrease towards the periphery of the lung. The deposition of particles within the respiratory tract depends on two major mechanisms – inertial impaction and gravitational sedimentation. Inertial impaction can be a problem if the airstream is moving rapidly, e.g. after pressurised aerosol release and/or a sharp inhalation. Thus, given excess force, the particles may simply impact on the walls of the airway and never reach the desired site in the lung. Alternatively turbulence can occur if the inhaled flow rate is too low. This may also prevent particles from reaching the desired site.

When the technical difficulties of aerosol generation, flow rate, etc., have been resolved, it becomes possible to treat experimental animals with the drug. With rodents this can be achieved by putting the animals, housed in cages, into large chambers so that the whole body is exposed to the material. This has the advantage of exposing large numbers of small animals simultaneously and obviates the need for restraint. Such a treatment method is, however, very wasteful of drug and the animals are exposed to skin absorption and mouth ingestion, both directly and by grooming drug from fur, and in prolonged studies food or water contamination may occur. Animals also tend to reduce exposure to noxious aerosols, by huddling together or by covering their noses, while loss of drug on the walls of the chamber can also occur.

It is therefore more usual to use a head (snout) only exposure. In such a system the animal is restrained in a plastic tube with only its nose protruding from one end into an inhalation chamber. Rodents are obligatory nose breathers and thus nose only exposure ensures that the animal will respire the aerosol or vapour. Most larger laboratory animals can breathe through their mouth or nose and thus nose-only methods are inappropriate. Instead masks may be used which cover the mouth and nose but prevent material from affecting the eyes. An alternative method for larger animals is to use oropharyngeal or intranasal cannulae, although the latter method may require anaesthesia thus altering normal airflow. The oropharyngeal route bypasses the nasal sinuses, thus permitting larger particles to reach the bronchii.

A major problem with exposure by inhalation in nose and mouth breathers is that only a small proportion of the dose actually reaches the alveoli, with only slightly more reaching the conducting airways, while the majority is swallowed. In spite of these considerations, the small amount of drug inhaled can provide a full therapeutic effect. The swallowed drug will, however, pass into the stomach and can then be absorbed from the gastrointestinal tract; a factor which may be of significance if the compound is orally active. Such an effect can also produce problems when attempting to measure blood drug levels produced by inhalation only.

Regulatory guidelines, though not standardised internationally, are broadly defined with regard to inhalation toxicity testing. The following general points arise from the guidelines. Inhalation toxicological studies should be undertaken in two species (one non-rodent) using the formulation which is to be marketed. All guidelines suggest that local as well as systemic effects be examined, while some (e.g. UK and EEC) suggest that parameters such as effects on mucociliary activity be examined. (To examine systemic effects, as described previously, it is possible to use routes of administration other than inhalation.)

It is also necessary to show that the techniques used in the inhalation studies actually do deliver the test substances to the desired site. Thus it is necessary to have information on the physical characteristics of the study, i.e. characterisation of particle size distribution in the aerosol, measurements of the atmosphere present in the exposure chamber (to ensure aggregation is not occurring), chamber humidity, temperature, air flow, etc., as well as biological measurements. These include methods described previously, but with special attention to measurements of respiratory physiology parameters and histopathology of the lung and respiratory tract.

It may also be necessary to determine whether there are any pharmacokinetic or metabolic effects relevant to the toxicological findings resulting from inhalation. This may be important if the drug is extensively metabolised by the lungs or if treatment causes enzyme induction, etc. As stated, above, interpreting blood drug levels may be a problem, owing to the material being swallowed and absorbed through the gut. Direct tracheal administration of the drug may assist in giving more meaningful metabolic and pharmacokinetic data.

The duration of inhalation toxicity studies, in two species, necessary to conduct clinical trials are similar to those required for drugs given by other routes. As is the case with most toxicity investigations, the first study is to determine the acute toxicity of the material (approximate median lethal dose) in rodents (usually rats). Groups of rats, of both sexes, are exposed to an atmosphere containing the compound, with the duration of exposure varying from several minutes to several hours. In most instances, with dose being dependent upon concentration and/or time of exposure, the time of administration is kept constant and the concentration of drug varied. In subsequent studies, groups of experimental animals (three dose groups and a control, the 'control' may consist of two groups – one treated with the vehicle (propellant), with the other untreated) are exposed to the drug at multiples of the therapeutic concentration, with the top dose usually expected to show a 'toxicological effect'.

Several distinctive features are inherent in these guidelines. In acute studies the characterised aerosol is delivered at a high concentration (up to 5 mg/l of inspired air if possible) consistent with the physical integrity of the aerosol. The treated animals must be kept after the exposure for a minimum of 14 days to allow the development of delayed clinical/pathological effects. Long term toxicity studies may also incorporate a postdose recovery period to allow the development of any delayed effects.

7.4 OCULAR

The albino rabbit is the species most commonly used in ocular toxicity studies. Reasons for this include the fact that rabbit eyes resemble mans in terms of humour volume and in general size and shape, but perhaps most important is the fact that rabbits have 'loose-fitting' lower lids which can be used as 'sacs' for instillation of test materials. In other species where the lower conjunctival sac is smaller or missing the drug can be directly installed onto the cornea ensuring, naturally, that the cornea receives the entire dose. Drugs can also be placed in the eye using ointments (these may, however, blur the vision, resulting in attempts to wipe the material away) or a suitable gel. Other more sophisticated methods include the use of soft contact lenses saturated with drug.

While the duration of ocular toxicity studies is dependent upon the proposed length of human exposure, it is unlikely that studies longer than 90 days would be considered. As in all toxicity studies, treatment should be carried out in at least two species and include one or two treatments which exceed the human dose, thereby demonstrating probable margins of safety. In acute studies it is usual to treat only one eye leaving the other to act as an untreated control, however, for subchronic exposures it may be advisable to include a vehicle control group.

The eyes of the experimental animals are examined closely before (to ensure they are free of defects), during and after (possibly including a 'recovery' group) treatment. Examinations include general and specialised ophthalmic observations, e.g. the slit lamp technique. The slit lamp can be used to detect a variety of changes accurately, e.g. thickening of the cornea, iritis, changes in the aqueous humour, etc. Such measurements, however, require skilled experimental personnel. Fluorescent dyes such as sodium fluorescein can also be helpful in detecting defects in the surface epithelium of the cornea and conjunctiva. The usual procedure is to place a couple of drops of dye (0.25 to 1.0% in sterile ophthalmologic solution) onto the cornea of the experimental animal. The dye is then washed off and the eyes examined by slit lamp technique with the damaged areas, which absorb the dye, fluorescing in the bright light. At the end of dosing the animals are killed and the eyes removed for histological examination. Buffered formalin can be used to 'fix' the eyes, but while fixation of the cornea, conjunctiva and extraocular tissues will be adequate, the intraocular tissues (lens, retina and iris) may not be ideally preserved. It is therefore better to use Zenker's or Davidson's fixative.

7.5 DERMAL

Dermal toxicity studies, usually undertaken with compounds designed for topical application to the skin, should preferably use the drug in its clinical form. Although a variety of species can be used in such studies, the adult albino rabbit is perhaps the most common. Reasons for this include size, ease of handling and because its skin is reported as being the most permeable of the common laboratory animals. A problem with the rabbit, however, is its extreme sensitivity to dermal insult, which can result in reactions not valid in man. To overcome this 'sensitivity' a second species such as the rat may be included; this has the added advantage of being the species which is generally used in systemic toxicity studies.

The usual procedure is to shave the fur to expose not less than 10% of the body surface over which the test material is applied. With liquids or solvents the agent may be held in place with a porous gauze dressing, but in normal circumstances occlusive dressings are not used since such coverings can enhance penetration thereby increasing potential toxicity. A problem with leaving the treatment 'open', however, is that the animals may ingest the test material. To prevent this, animals should be housed singly and some sort of screen or other device used to cover the applied material. The basic design of dermal toxicity studies, which may last for up to 90 days of repeat dosing, is similar to that used for other types of toxicological investigation, i.e. three dose levels and a placebo control, use of rodent and non-rodent species, group size increasing with increasing duration of treatment, etc.

At the end of the treatment procedure the animals are given a thorough examination, especially the exposed area of skin, before they are killed and subjected to a *post mortem* examination. The only tissues taken at necropsy are skin samples from the treated areas and any other areas showing abnormalities. If treatment is associated with skin irritation it may be worth while to include a 'recovery group' to determine whether or not the effects are reversible.

An important part of dermal studies is the measurement of percutaneous penetration. This is achieved by measuring concentrations of drug and/or metabolite(s) appearing in the blood, urine, faeces, etc., following dermal application. These studies can be performed following single or multiple dosing but, since the skin can retain chemicals for relatively long periods of time, examination of body fluids for drug metabolism and pharmacokinetics should last for a minimum of five days. If the drug is found to be distributed around the body, it will be necessary to perform systemic toxicity studies using intravenous or oral dosing.

7.6 RECTAL

Rectal administration is seldom undertaken in species other than dogs and rabbits. Dogs have the advantages of being able to accept human suppositories (two at a time if necessary), they can be dosed three or four times daily to achieve multiples of human dosing and are sufficiently robust to permit long term studies. Although rabbits are also useful for such studies, it is rare to dose them for longer than 30 days. In all studies involving rectal administration, the animals should be observed closely for up to an hour after dosing. If the medication is quickly rejected (within 20–30 minutes), it is usual to repeat the administration.

7.7 VAGINAL

Rabbits and dogs are the species used for this route. In rabbits studies seldom exceed 30 days' duration, but longer studies are possible in dogs. Dogs also have the advantage that human vaginal pessaries can be administered. Both species can be used to evaluate creams and ointments.

7.8 SUBCUTANEOUS

In rodents such studies rarely exceed 30 days' duration. Injections are made just below the skin of the flank with a 25 G, 15 mm needle, with alternate sides of the animal being used each day. In dogs the injection is made into the loose skin above the shoulder blade, which can be raised with the thumb and fingers. Pointing the needle downwards and towards the fingers avoids the possibility of the needle going straight through the skin fold.

7.9 INTRAMUSCULAR

Owing to the small size of the mouse, it is not feasible to conduct intramuscular injection in this species for longer than a week. Other laboratory rodents such as the rat can be used for 30 days, while primates and dogs may be dosed for longer periods. For intramuscular injection in rats two operatives are usually needed, one to restrain the animal and the other to hold the leg and make the injection. The *biceps femoris* is normally used and it is important to avoid the sciatic nerve. The maximum volumes which can be given to a mouse and a rat are about 0.2 and 0.5 ml respectively, using a 25 G, 15 mm needle.

When dosing dogs it is usual to stand the animal on a table, ensuring that it is firmly held to prevent any sudden movements. The hind legs are shaved to give a clear view of the *quadriceps femoris* muscle, and the target area is swabbed clean. The appropriate dose is given using a sterile disposable needle which is pushed 2 – 3 cm into the muscle. If the muscle is accurately dosed no damage will occur to the sciatic nerve.

7.10 INTRAPERITONEAL

To give drugs by intraperitoneal dosing, the animal is laid on its back and the abdomen shaved. This area is thoroughly cleansed and, using an appropriate syringe and needle, the abdominal wall is punctured. To ensure minimal danger of perforation of abdominal viscera, the injection should be made rostral and lateral to the bladder at an angle of about 15° to the abdomen. The depth of penetration should not exceed 5 mm.

There are a variety of other sites at which it may occasionally be necessary to dose animals in order to provide information about local toxicity. These include nasal sinuses, joint capsules, intraspinal injections, etc. Such methods of dosing are not common and in the case of the last, cannot be repeated for prolonged periods of time.

REFERENCES

Willeke, K., Lo, C. S. K. & Whitby, K. J. (1974) Dispension characteristics of a fluidised bed. *Journal of Aerosol Science*, 5, 449–55.

8

Reproductive toxicology

The ability of chemicals to produce foetal abnormalities has been known for some years, but such effects were considered unlikely to be produced at therapeutic dose levels of drugs. The fallacy of this belief was sadly demonstrated by the thalidomide tragedy, when it became evident that agents lacking 'toxicity' in toxicological studies could still possess marked teratogenic properties (i.e. capable of causing birth defects). Regulatory authorities and pharmaceutical companies were then faced with a problem which required prompt action to ensure other embryo and/or foetal toxic materials were not used in the human population. Since the studies then in use were inadequate for detecting potential developmental toxicants it became necessary to introduce methods to detect those agents capable of affecting normal reproductive processes.

There was some criticism that, to begin with, the major focus of attention was on the detection of malformations (teratology), with less regard being paid to other aspects of reproductive toxicity. In later years, however, reproductive studies have been designed to study not only teratological potential, but also gametogenesis (in both sexes), mating behaviour and performance, fertilisation, implantation and embryogenesis. The length of gestation, parturition, lactation, weaning, maternal care and a variety of postnatal effects, including growth, learning ability, reproductive capacity, etc., are also examined. (Because of the ongoing debate concerning the use of different terminologies in reproductive toxicology, the authors have attempted to use only a limited number of terms in a simplified context.)

8.1 TYPES OF STUDY

The FDA in the United States was the first regulatory authority to issue specific guidelines designed to evaluate new pharmaceutical preparations for reproductive toxicology. During the past few years, however, a number of other countries, including Australia, UK, Sweden, Canada, Japan, and agencies, e.g. CPMP, have also established guidelines for reproductive studies. The major problem for the international pharmaceutical company is, therefore, to perform such tests which 'cover individual distinctions' but meet 'worldwide requirements'. In practice this can be quite complex since although without exception individual guidelines are consistent in demanding tests which can be categorised into three segments:

Segment I	Fertility study	
Segment II	Teratology study	see Fig. 8.1.
Segment III	Peri and postnatal study	

the actual content of each segment can vary quite considerably between authorities; this being especially true of Segment I.

Another difficulty in describing 'standard reprotoxicity studies' is that different laboratories working to the same regulatory guidelines can, and often do, use completely different protocols. Thus there may be differences both between and within regulatory study designs. For this reason some basic study protocols used in the UK will be described and, where critical variations between regulatory authorities occur, these will be described. It should be remembered that, owing to the pharmacological activity of a drug, it may not be possible to follow standard protocols, e.g. it is not possible to conduct a

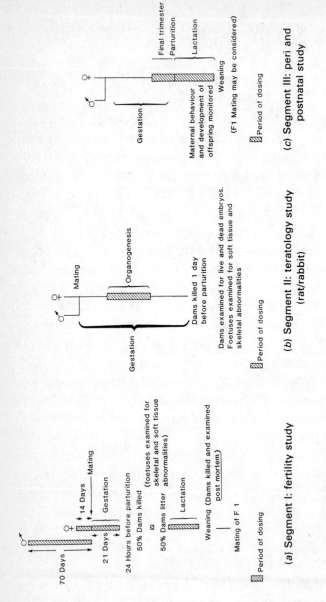

Fig. 8.1. *Standard reprotoxicity studies.*

(a) Segment I: fertility study

70 Days

14 Days

Mating

21 Days

Gestation

24 Hours before parturition

50% Dams killed

(foetuses examined for skeletal and soft tissue abnormalities)

50% Dams litter examined post mortem

Lactation

Weaning (Dams killed and examined post mortem)

Mating of F 1

Period of dosing

(b) Segment II: teratology study (rat/rabbit)

Mating

Organogenesis

Gestation

Dams killed 1 day before parturition

Dams examined for live and dead embryos. Foetuses examined for soft tissue and skeletal abnormalities

Period of dosing

(c) Segment III: peri and postnatal study

Final trimester

Parturition

Lactation

Gestation

Maternal behaviour and development of offspring monitored

Weaning

(F1 Mating may be considered)

Period of dosing

conventional fertility study using contraceptive drugs and/or other compounds directly affecting the reproductive process.

Before describing the procedures used in reproductive toxicology, it is perhaps opportune to remind the reader how essential it is, when conducting such studies, to pay very close attention to the times of dosing. For example, in Segment II studies the dam must be treated during the period of organogenesis, otherwise positive teratogens may well be missed. For any study, it is essential to be precise about the developmental stage of the embryo and the physiological state of the mother.

8.2 FERTILITY STUDY (SEGMENT I)

The fertility study has been designed principally to provide an overall screen for the effect of pharmaceuticals on gametogenesis, mating and fertilisation. Such studies, often called fertility tests, are almost always conducted in the rat, but there is no reason why other species such as the mouse cannot be used.

An important component of the fertility study is the examination of drug effects on male reproduction (the other two studies, i.e. Segments II and III, do not include male dosing). While the necessity for examination of fertility and reproduction in females is obvious to most people, it is often overlooked that many chemicals can affect male fertility, with adverse responses being brought about in a number of ways by a variety of agents. Examples of the ways in which agents can affect fertility and reproduction in the male include (*a*) direct effects on the sperm causing sperm death, reduced sperm count, abnormal sperm or mutation effects, etc. (e.g. drugs like thiotepa and cyclophosphamide), (*b*) testicular hypertrophy (e.g. strong anti-androgens) and (*c*) decreased libido caused by direct effect on the CNS, endocrine imbalance or phamacological effects (e.g. anti-cholinergics, which affect erection).

Segment I studies are often divided into male fertility (Segment IA) and female fertility (Segment IB) studies. In the Segment IA study male rats, six to seven weeks of age at the beginning of treatment, are randomly allocated to treatment (usually three dose levels) and control (vehicle treated) groups. Each group, usually containing at least 20 animals, is treated for 70 days prior to mating, usually on a one to one basis, with untreated females (dosing is continued during mating). This treatment regime ensures that the animals are exposed to the drug throughout the whole cycle of spermatogenesis, including storage and maturation of spermatozoa. During mating the females are examined daily for evidence of a vaginal plug or presence of sperm, which would indicate successful copulation had taken place. If copulation does not occur, the male is paired with another female which is again examined daily for the presence of a vaginal plug. Success rates of mating and fertilisation are recorded. If fertilisation does not occur, samples of sperm from the treated males may be examined for sperm count, motility and morphology.

The pregnant females are maintained until the penultimate day of pregnancy when they are killed, usually by CO_2 asphyxiation, and the young delivered by Caesarean section. The ovaries and uterus are kept for a subsequent count of the *corpora lutea* and determination of implantation sites. The position *in utero* of the live foetuses and resorption sites are noted, the latter being divided into early and late embryonic deaths, i.e. inability and ability to distinguish between foetal and placental tissue respectively. The live young are weighed, sexed and subjected to a thorough external examination before being killed by pentobarbitone overdose. It is unusual, on a male fertility study, to perform any further examination of the pups, but it is possible to keep them for both visceral and skeletal examination (see teratology studies).

The males are usually maintained on treatment up to the point when the pregnant females are killed; this being an 'insurance policy' so that in the event of the death of a pregnant female the male could be remated.

The males are finally killed and a macroscopic examination of the reproductive tissues is undertaken. Microscopic examination of these tissues is not usually carried out, since this is included in the other toxicity studies (see Chapter 4).

A major weakness of the male fertility study is that the rat has such an excessive production of sperm that a large proportion (up to perhaps 90%) may be adversely affected without reducing fertility. In contrast the human male is 'almost infertile' and a slight effect on the sperm count can render him infertile. Thus it has been argued that the rat is an insensitive species for studying male fertility and, instead of using successful mating as an end point, other functional tests, such as the ones described previously, e.g. sperm count, should be used. Reproductive toxicologists are also debating whether a 70 day dosing regime is necessary, since it is argued that the majority of damage occurs in the late development of sperm. Thus it has been suggested that dosing could be reduced to about a quarter of its present level.

In the Segment 1B study female rats, housed individually, are randomly allocated to three treatment groups and control groups (30 animals per group). Dosing begins at least 14 days before mating (this is an attempt to cover at least three oestrous cycles; a normal cycle in the rat lasts four days) and continues daily until the animals are killed. Vaginal swabs are also taken daily, prior to mating, to assess ovulation and normal cycling. Once pregnancies are assured, the females are again housed individually with daily monitoring for general health and body weight changes. Half the dams are killed before parturition with the remainder being allowed to litter and wean their young.

For those animals killed before parturition there are some differences in the regulatory requirements regarding the time at which the dams should be killed and the type of analysis required on the foetuses. With regard to the timing of death, the USA and Canada prefer mid-pregnancy kills (although the FDA will accept term killing), while the Japanese require them just prior to parturition. Other guidelines such as the UK, CPMP and Australia simply require the dams to be killed some days before parturition. In order to meet the varied requirements some protocols have included large group sizes (e.g. 45 females) so that adequate numbers of dams could be killed mid-term and at the end of term, with the remainder being allowed to litter, i.e. 15 animals per time point. It is now more usual to kill the rats just before parturition and omit the mid-pregnancy kill.

Another regulatory issue is the type of assessment which should be performed on the killed animals. For a 'fertility study' it would be expected that an assessment of number of pregnancies, preimplantation and postimplantation deaths, together with a careful macroscopic examination of the foetuses, etc. would be adequate for the authorities. The Japanese, however, require a full visceral and skeletal examination of the foetuses similar, if not identical, to those performed in teratology (Segment II) studies (see Appendix 1).

It is now becoming standard practice, for both scientific and pragmatic reasons, to include such examinations in all female fertility studies. Scientific reasons include the production of further data which can be included with those produced in teratology studies, making total group sizes of 40 rather than 25 females. Also, when performing a 'combined fertility study' (treated males mated with treated females), any 'male mediated' effects on foetal development may be detected. The pragmatic reason is the unhindered marketing of drugs in Japan.

As the procedures used for examination of the foetuses in a fertility study are identical to those used in a teratology study they will not be described here. A full description of such techniques is given later in this chapter (see pp. 127–128). The remainder of this section will therefore concentrate on the examination of the offspring.

The dams which are allowed to survive to the end of gestation are allowed to litter, often with constant observation throughout parturition. Since this usually occurs at night, fertility studies are sometimes run on a reversed night and day cycle using dim red lighting during the 'dark' hours. At birth the numbers and sexes of the live and dead pups are recorded, with the surviving animals being closely examined for any variations from normal behavioural and physical development. Especially important development markers are changes in body weight, hair growth, tooth eruption, opening of eyes and ears and the acquisition of various reflexes, e.g. response of individual pups to light and of whole litters to sound. Maternal behaviour is also closely monitored to ensure correct maternal care, weaning, etc. Once the pups have been weaned, the dams can then be killed and the abdominal and thoracic viscera examined for gross abnormalities.

Close attention is also paid to the reproductive capacity of the sexually mature offspring, with a single male and single female from each of the 15 litters being examined. In such studies males and females, avoiding brother–sister mating, are paired individually (female transferred to male cage) overnight. The next morning the animals are checked for proof of mating, signified by a positive vaginal smear and the presence of a copulation plug in the vagina or on the floor of the cage. If no copulation plug can be found the female is given a vaginal smear by gently flushing the vagina with sterile isotonic saline and placing all deposits on a clean dry microscopy slide. The slide is then examined by low power microscopy for the presence of spermatozoa and to determine the stage of oestrus. If any pair fail to mate within 15 days the female will be mated with another F1 male, of proven mating ability, and the male mated with another female.

The Japanese Segment IB design is radically different from

that described above. In the Japanese study, dosing of females stops on pregnancy day seven and all females are killed at term, i.e. there are no offspring. It is, however, unusual for Western companies to conduct female fertility studies using this design.

The considerable quantity of data generated by such a study is analysed statistically to reveal any treatment-related differences in reproductive capacity, ovulation, implantation, intrauterine survival and development, nature and incidence of foetal abnormalities, maternal and pup weight gain and in postnatal survival and development.

It is now becoming usual to mate dosed males with dosed females ('combined study'), thus reducing the number of animals which must be used. Some countries, however, such as Australia, require that the dosed animals be mated with undosed ones ('separate study'). It is not clear what advantages the 'separate study' has over the 'combined', except of course if an effect is seen in the latter and it is not possible to determine which sex has been affected without repeating the work using separate sex studies. The Australian authorities are expected to take a flexible approach, assuming that whichever protocol is followed the study is performed to a high standard. It is also possible to conduct fertility studies using male animals from a conventional three or six month toxicity study. This latter approach may produce logistic problems and obviously the animals are not killed after successful mating.

8.3 TERATOLOGY STUDY (SEGMENT II)

Teratology studies are mainly concerned with detecting those agents that can directly or indirectly damage the embryo or foetus at doses which may not necessarily produce any toxic responses in the mother. The importance of these studies lies in the fact that the action of an agent on the foetus is very often irreversible, and such foetal damage can lead to morphological or functional abnormalities in the newborn. In order to be able to detect such effects the test agent is dosed to the pregnant female over a specific part of pregnancy, i.e. the period of organogenesis when the embryo is at its most vulnerable to the action of teratogens.

It is usual to use two species for this type of study (guidelines state a rodent and non-rodent species). Rats and rabbits are by far the most common, although for certain compounds other species ranging from mice to monkeys have been used.

In the rat, groups of at least 25 adult females have vaginal smears checked daily for at least two oestrous cycles to confirm that ovulation is normal. They are then mated with adult males which have previously been demonstrated to be fertile. Pregnancy is considered to begin on the first day in which a vaginal plug is found. The pregnant animals are then randomly allocated to treatment groups (usually three dose levels and a placebo control) with treatment lasting from days P6 to P15 (P7 to P17 in Japan) (See Table 2.3 and Appendix 1), i.e. to cover the period of organogenesis. On the penultimate day of gestation the dams are killed, the uterus exposed and the foetuses delivered by Caesarian section (killing the dams prior to expected parturition overcomes the problem of malformed or sickly offspring being devoured by the mother).

The numbers of viable and dead embryos are counted to quantify intrauterine deaths. Early and late embryonic deaths, again being characterised by the ability to differentiate or not between foetal and placental tissue, are measured (length and diameter) together with the site of implantation. Each foetus is then examined externally for gross abnormalities such as anencephaly (congenital absence of the brain) or umbilical hernia. If foetuses are alive they are weighed and measured before being killed with a barbiturate overdose. Alternate pups are selected for 'fresh visceral' examination and subsequent skeletal preparation, or immediately fixed in Bouin's fluid for subsequent detailed examination of internal organs. It has been suggested that one third of the foetuses should be used for organ examination and two thirds for skeletal. However, since there is no real scientific evidence that the skeleton is more susceptible to teratogens than the other organs, the 50:50 distribution seems reasonable.

There are two major methods for performing a visceral examination. The first (Wilson 1965) involves freezing the fixed pup in solid CO_2 and then, using a razor-blade, cutting (free-hand) transverse sections through the body. The individual sections may be stained, e.g. with cresyl violet, before being examined by an experienced teratologist. The sliced, fixed embryos are kept as raw data. In the second method, the so-called *in situ* section technique, the foetuses are examined using standard necropsy procedures. It has been suggested that because the structures and organs are in their more normally recognisable positions the technique is easier to learn and perform. Also desired organs may be removed and subjected to further examination by longitudinal and/or transverse sectioning.

For skeletal examinations the foetuses are eviscerated (the internal organs being examined and discarded) skinned, fixed in alcohol, processed through alcian blue (cartilage stain), alcohol, potassium hydroxide and finally stained in alizarin red S stain. The stained skeletons are then placed in a solution of 70% alcohol and glycerin (1:1). These permanent records can then be examined and, if any problems occur, can be rechecked for a second opinion.

Rabbit studies usually employ smaller group sizes, 13–15 being typical. Primiparous does, possibly primed by a gonadotrophin injection, are mated with or artificially inseminated by a proven donor buck. Such 'time mated' females are supplied by several animal breeding companies. Treatment of the pregnant rabbits is from day 8 to 18 of gestation and they are killed on day 28. Similar examinations are performed to those for rats.

Some authorities including Australia, but more especially Japan, are interested in using teratology studies to detect not only agents with the potential to cause foetal death, structural malformation, etc., but also those which may affect postnatal development. Thus it is necessary not only to determine effects on the unborn, but also to examine for any delayed effects which may only become apparent during the development of the offspring. In variations on these teratology studies, an extra 10 females per group are allocated and allowed to litter normally. Their offspring are observed until they are 42 days old, to monitor physical and behavioural development as in fertility studies. Some of these offspring may also be mated, as in a Segment IB study, to measure reproductive capacity. This enables some assessment of small, drug-induced, embryological changes which may not be apparent on examination of the foetuses taken from their mothers before parturition.

Although it is possible to expand a Segment II study to include observations of the offspring, there is some debate as to whether such an extended teratology study is really necessary, especially in view of the other two reproductive studies (Segment I and III) which are also performed.

8.4 PERI AND POSTNATAL STUDY (SEGMENT III)

This test, almost always performed in rats, is perhaps the simplest of the reproductive studies. Its objective is to investigate possible adverse effects on late intrauterine growth, parturition and postnatal maturation. The study is also designed to detect agents which may cause problems with labour and/or delivery, lactation (i.e. inhibit milk production), milk ejection or produce effects in the offspring due to delivery of the drug to the pups via the milk.

Groups of 20 females are mated as described earlier (p. 123). The pregnant animals are then dosed throughout the final third of gestation, i.e. from day 15 (day 17 in Japan), until the young are weaned at day 21 *post partum*. The animals are allowed to litter and the young periodically checked until weaning. During this period of the study a variety of assessments is made, e.g. labour, delivery, number of live and dead young, litter size, malformations, neonatal viability, lactation. Under certain circumstances (e.g. if results of Segment I and II studies indicate possible problems) some of the progeny may be allowed to mate so as to assess their reproductive capacity. (In Japanese guidelines such offspring mating assessments are mandatory.)

In recent years it has become considered to be important to examine progeny not only for physical abnormalities, but also for behavioural impairment. A problem with behavioural reproductive toxicology is that although regulatory authorities such as the EEC, OECD, UK, Canada and Japan require that the progeny be examined for such effects, they are often not precise about the nature of the studies which should be undertaken. It has been suggested that pups from Segment III studies, in which the foetuses are exposed to drugs in the last third of gestation (i.e. at the time they are most vulnerable to CNS effects), should be used for examination of behavioural changes.

As stated above, although test methods for behaviour observations on pups have not been specified, investigations are at present being performed in several countries to establish testing systems. For example, the European Office of the World Health Organisation published some draft guidelines for assessing drugs for behavioural teratogenicity in 1986. However, although the proposals included reflection (righting on a surface), movement, sense, activity, learning and memory as items for examination, no descriptions of definite methods were given. The US Environmental Protection Agency (1988) has also published guidelines for behavioural teratogenicity studies on specified chemicals such as diethylene glycol, butyl ether and diethylene glycol butyl ether acetate. At present, with the lack of any clear recommendation, behaviour observation studies should be employed according to the character of the drug and the proposed patient group that will receive it. Other references on behavioural teratology which may prove useful to the reader include Tanimura (1985) and Riley & Vorhees (1986).

8.5 DISCUSSION

It is generally assumed that the major cause for concern to authorities, pharmaceutical companies, etc., is the ability of chemicals to cause teratogenic effects. From the previous pages it has been demonstrated that studies are conducted which cover all aspects of reproductive toxicology, not just teratology.

An important aspect of reproductive toxicology, however, is that the observed effect of a chemical is very much dependent upon a number of variables, such as species in which testing is performed, dose of chemical and stage of reproductive cycle or gestation at the time of administration.

A good example of how different species show a different susceptibility to the teratogenic potential of a chemical can be seen with thalidomide. Whereas rats are relatively resistant and mice show some reactions (not usually limb effects), rabbits, monkeys and man are highly sensitive to the teratogenic action of this agent. It is for this reason that the regulatory authorities demand that, for teratogenic (Segment II) studies, test materials be examined in at least two species. For teratology studies the rat and the rabbit are the usual species chosen although others include the monkey, hamster, pig and ferret.

Of interest in the selection of species for teratology studies is the investigation, reported by the FDA in 1980, in which they examined the ability of different species to identify correctly agents known to be teratogenic or non-teratogenic in man.

The study, which used five species (i.e. mouse, rat, rabbit, hamster and monkey), included 35 positive and 165 negative human teratogens. The order of success for the correct identification of the positive agents was mouse (correctly identifying 85%), rat (80%), rabbit (60%), hamster (45%) and monkey (30%). The mouse and the rat were therefore the most sensitive of the species examined, with the monkey being the most insensitive. However, for correctly identifying true negatives (a measure of specificity), the order was almost reversed with the monkey correctly identifying 80% of the negatives, then the rabbit (70%), rat (50%), hamster (35%) and mouse (35%).

In terms of sensitivity, although the mouse detected most teratogens it also produced many false positives. The monkey, while showing good specificity, can be criticised for lack of sensitivity. It is possible, by performing a simple calculation, to derive a rough value for the accuracy shown by each of the species:

$$\text{accuracy} = \frac{\text{no. of correct results}}{\text{no. of chemicals tested}}$$

Using this simple formula the accuracy for each species is approximately: monkey 70%, rabbit 69%, rat 56%, mouse 44% and hamster 34%. Thus, in terms of accuracy, the two species routinely used, i.e. rat and rabbit, would appear to be the best choices although both (especially the rat) show a tendency to overestimate the teratogenicity of chemicals. (Monkeys are not routinely used in these studies.) For the other studies, i.e. Segments I and III, the rat is the most commonly used species although the mouse may occasionally be used.

Dose selection is an important factor for the pharmaceutical company, since it is possible that developmental toxicity may occur at a treatment many times greater than the dose level intended for use in man. Thus there may be some problems in selecting 'sensible' doses for use in reproductive studies. Before undertaking reproductive studies it is usual to perform limited dose ranging studies in pregnant animals, since data on maximum tolerated doses from non-pregnant animals may be misleading when extrapolated to the pregnant animal.

In teratology studies the highest dose should cause demonstrable, but not severe, maternal toxicity. It is difficult to be precise about what represents demonstrable toxicity, but a rough guide is that treatment should not reduce body weight gain in pregnant animals, as compared to controls, by more than 10%. If doses are used which produce greater retardations in weight gain, this may result in reduced foetal body weight which can cause an increased incidence of minor malformations. If adverse effects in the conceptus are only seen at doses producing severe maternal toxicity, then it can be assumed that the test agent is not exerting a selective developmental toxicological effect.

For Segment 1B (female fertility) studies it is desirable to use a treatment which reduces foetal or maternal birth weight by about 10 to 15% at the high dose (this information should be available from the rat teratology study or a sighting study). A greater reduction in body weight can have adverse and irreversible effects on postnatal growth, survival and behavioural development. In Segment IA (male fertility) studies it is usual to use doses similar to those used in subchronic toxicity studies.

An important factor when dosing the mother is to determine whether the drug is able to cross the placental barrier, because accessibility of the drug to the foetus may be the reason for variation in species susceptibility. It is therefore usual to conduct a study in which pregnant rats are dosed daily with the drug and drug levels are measured on two occasions in maternal plasma, amniotic fluid and foetal tissues. Another method is to use whole body autoradiography of pregnant experimental animals treated with radiolabelled drug and look for activity in the foetuses. If there is little or no placental transfer it may be necessary to select another species or find an alternative method of assessing embryo toxicity. Of course teratogenic effects can occur indirectly, thus the absence of a drug or metabolite in foetal tissue does not necessarily prove safety.

Pharmaceutical companies are concerned with selection of the optimum time in the drug development process at which to perform reproductive toxicity studies. If performed too early, resources may be squandered on large and complex studies with compounds not destined to stay in development; alternatively, important clinical trials should not be delayed because essential studies have not been performed. While in the majority of instances it is necessary to have completed and reported all three reproductive studies prior to drug registration, it is more difficult to be precise about the requirements for such studies in order to carry out clinical trials. For instance in Nordic countries the guidelines describing requirements necessary to undertake clinical trials make no mention of reproductive toxicity studies (see Appendix 1), while in the UK teratology (Segment II) studies are required before women of childbearing age can be included in clinical trials. Other reproductive studies are not, however, required until a product licence application is made.

In the USA, before carrying out clinical trials in women, it is first necessary to have a two-species teratology study, a female fertility study and some human clinical data. The full panoply of studies is, however, required for Phase III clinical trials (see Table 2.5 and Appendix 1). Similarly, in Australia teratology studies are required prior to starting Phase II trials. In Canada, while no reproductive studies are required for Phase I trials, they are necessary before undertaking Phases II and III.

Because of such regulations, companies with too little reproductive toxicity data may find themselves restricted to only a limited number of countries in which they can perform clinical trials. Naturally it is always possible to conduct preliminary studies in men or in women past childbearing age.

REFERENCES

Riley, E. P. & Vorhees, C. V. (1986) *Handbook of behavioural teratology.* Plenum Press, New York.

Tanimura, T. (1985) Guidelines for developmental toxicity testing of chemicals in Japan. *Neurobehaviour Toxicology and Teratology,* 7, 647–52.

US Environmental Protection Agency (1988) Diethylene glycol butyl ether and diethylene glycol butyl ether acetate; test standards and requirements. *Federal Register,* 53, 5932–53.

US FDA (1980) *Federal Register,* 45, (205), 69823–24.

WHO (1986) *Guidelines for the assessment of drugs and other chemicals for behavioural teratogenicity* (draft). World Health Organisation, Regional Office for Europe.

Wilson, J. G. (1965) Methods for administering agents and detecting malformations in experimental animals. In *Teratology Principles and Techniques,* ed. J. G. Wilson & J. Warkanay, University of Chicago Press, Chicago and London. pp. 262–77.

9

Genotoxicity

The purpose of this chapter is to introduce the reader to genotoxicity (mutagenicity) testing, provide a description of the major test systems currently in use, speculate on future developments and describe the regulatory requirements associated with this area of toxicology. It is not, however, possible to cover this topic without examining some of the reasons why genetic toxicology has, in a relatively short period of time, achieved such apparent importance and why, in comparison to other branches of toxicology, it generates such varied and extreme opinions.

Before considering the role of genetic toxicology in the pharmaceutical industry, it is important to understand the rationale for undertaking such studies and the reason why such importance is attributed to them. Genotoxicity testing is based upon the knowledge that various agents, collectively known as mutagens, possess the ability to interact with and damage DNA. Such DNA damage occurring in somatic cells may result in tumour development, while abnormal offspring could result from affected germ cells. The obvious requirement, therefore, is to prevent individuals from being exposed to genotoxic agents, i.e. chemicals which may not only produce mutations *per se*, but also have other effects on the genetic material. Thus, over the past 15 years or so, much effort has been expended in developing test systems that can be used to detect chemicals which possess the specialised property of interacting with and altering the hereditary component of living cells. The term genotoxicity has become popular for describing such events.

To date well over 100 assays have been described which, in theory at least, have the ability to detect potential carcinogens. Naturally it is not possible, in this text, to describe all such systems. Instead the principles of these tests, which are believed to be the most informative, predictive, best validated and most helpful in the legislative context, will be described.

For convenience the genotoxicity assays have been divided into the endpoints they measure viz. gene mutations, chromosomal effects, DNA interactions and neoplastic cell transformation.

9.1 GENE MUTATION ASSAYS

Bacterial test systems

Of the bacterial mutation assays available, the *Salmonella typhimurium* system developed by Professor B. N. Ames and his colleagues (Ames, McCann & Yamasaki 1975) is the one most widely used and is probably the system by which the largest numbers of chemicals have been evaluated. The basis of the test has been the development of mutant strains of *S. typhimurium* which have lost the ability to synthesise the essential amino acid histidine, and can therefore only grow and divide if an exogenous supply of the amino acid is provided in the culture medium. Agents producing mutations in the defective gene, either by qualitative or quantitative changes in the nucleotide component of the codon (i.e. base pair substitution or frame-shift mutation respectively), cause the bacteria to revert from being dependent on histidine for growth to a histidine independence. Thus any mutant cells grow and divide in a medium deficient in the amino acid, while non-mutated cells fail to form colonies.

As the gene defect producing histidine deficiency is known, it has proved possible to produce different strains of the bacteria to detect both base pair substitution and frameshift mutagens. (The strains designated TA 1535 and TA 100 are able to detect base pair mutagens, while TA 1537, TA 1538 and TA 98 are able to detect frameshifts.) The strains have also been manipulated to make them more sensitive to chemical mutagens; the cell wall has been altered to make the cell more permeable to chemicals, and they are deficient in their ability to repair certain types of DNA damage (see section 8.3, DNA interactions). Recently other strains of *S. typhimurium* have been introduced, e.g. TA 102, reputedly offering advantages in sensitivity over the other types. At the time of preparing this chapter the strains of bacteria most widely used are those described earlier, but it is possible that as time passes other strains may be introduced.

Most other bacterial mutation assays which have been developed also rely on the reversion from autotrophy to prototrophy, e.g. the *Escherichia coli* WP$_2$ series of bacteria which measures reversion from tryptophan dependence to tryptophan independence as the end point. It has been claimed (Bruisk 1980) that such assays have, however, no advantage over the Ames test and do not offer improvement in detection capabilities. Despite such comments, the Japanese regulatory authorities require that bacterial mutation assays are performed using not only *S. typhimurium*, but also *E. coli* strains.

Yeast test systems

A number of forward and reverse mutation induction assays have been developed using *Saccharomyces cerevisiae*. The most extensively studied system is probably the strain with defects of the genes for adenine synthesis. Forward mutation is detected by the production of colonies without the coloured pigment characterising the parent colonies. Both frameshift and base pair substitution mutations can also be studied in yeasts by measuring the reversion from autotrophy to prototrophy using a number of amino acids, e.g. reversion to methionine prototrophy in *cerevisiae* strains S138 and S211α.

Mammalian cell test systems

The major markers used in mammalian cell mutation assays are the detection of mutations at the hypoxanthine-guanine phosphoribosyl transferase (HGPRT) or thymidine kinase (TK) loci. The first of these assays is based on the fact that cultured cells which suffer mutations at the HGPRT locus, resulting in the loss of enzyme activity, can then grow and divide in a culture medium containing the normally cytotoxic purine analogues 8-azaguanine (8-AG) or 6-thioguanine (6-TG). Since the HGPRT system is essentially a salvage pathway, its loss does not result in any reduction in cell viability and therefore the mutated cells can be detected by counting the number of colonies.

Resistance to bromodeoxyuridine (BUdR) or trifluorothymidine (TFT) has been used to detect mutations in the mouse lymphoma cell line L5178Y. The specific cells used are heterozygous at the thymidine kinase locus, i.e. $TK^{+/-}$, so that they can undergo a simple step mutation to become $TK^{-/-}$. These mutants with little or no thymidine kinase enzyme activity are not killed when incubated with TFT or BUdR since they do not incorporate the toxic materials into their DNA.

Another mammalian cell mutation system which has received considerable publicity is the induction of ouabain resistance in cultured cells. By binding to the membrane bound Na^+/K^+ dependent ATP-ase, ouabain affects the sodium and potassium transport system, resulting in ionic imbalance and cell death. Thus normal 'non-mutated cells' are killed by adding ouabain to the culture medium, whereas the mutated cells, possessing an enzyme system which does not now bind ouabain, survive and grow in its presence.

Of the three *in vitro* mammalian bioassays described above, the most commonly used and widely advocated is the loss of HGPRT activity. There are a number of reasons for such popularity. First, it has the advantage over the other assays in the fact that the HGPRT gene, located on the X chromosome, is expressed on only one of the X chromosomes in the female cell (male cells naturally have only one X chromosome). Thus it is not necessary to use a heterozygous cell line, as is the case with TK activity. Second, the system has been widely used and exponents of such systems describe it as being the most sensitive. Another advantage is the fact that the method is relatively simple to perform and there are few problems in scoring 'mutant colonies'. In other assays, such as the L5178Y TK$^{+/-}$ system, there are complications in the production of 'large' and 'small' colonies, which have been described as representing 'point' and 'chromosomal' mutations respectively.

In all of the *in vitro* mutation assays described above, and indeed in the subsequent *in vitro* assays described later in this chapter, an exogenous metabolising system must be provided (major metabolic activation systems currently in use are

Table 9.1. *Types of metabolic activating systems commonly used in* in vitro *genotoxicity screening assays*

Activation system	Source
Post-mitochondrial (9000 g (S9) or 15000 g fraction)	From tissues (mainly liver) of rodents (mainly rats) previously exposed to agents which induce metabolic activation, e.g. aroclor 1254, phenobarbitone, etc.
Purified microsomes	As above
Freshly isolated cells	Mainly rat hepatocytes

shown in Table 9.1). Such an addition of 'metabolic activity' is essential when it is remembered that many genotoxins are not active *per se*, but require metabolising to an active form and most of the cells used in *in vitro* assays have little or no drug metabolising capability.

While *in vitro* mutation assays still dominate, a survey of genetic toxicology testing (Farrow, McCarroll & Anletta 1986) showed an overall increase in the use of both somatic and germinal cell mutation assays which employ a whole animal model. Such *in vivo* somatic cell mutation assays include tests using insects (the sex-linked recessive test in *Drosophila melanogaster*) and mammals (mouse spot test). Both of these test systems detect phenotypic changes in the offspring of treated parents. The advantages of such tests are that there is no need for an exogenous metabolising system (*Drosophila* are considered to possess many of the metabolising systems associated with mammalian liver), and true heritable genetic damage is assessed with many loci being monitored. There are of course many problems with such *in vivo* tests, not least being the relatively small number of chemicals tested in these systems (it has not been well proved), the breeding facilities required (knowledge of fruit fly husbandry or indeed dosing is not universal), and the predictive value of such heritable genetic damage to carcinogenicity being unknown. Indeed it is conceivable that certain chemicals, or their reactive metabolites, may never cross the placental barrier to reach target cells in the mammalian embryo.

It is evident that methods for detecting and analysing mutations *in vivo* are becoming more and more important, especially as dissatisfaction and cynicism is growing with regard to *in vitro* mammalian cell mutation assays. Attempts have been made to overcome the limitations of cell culture systems by developing procedures which can be used to detect mutations at the HGPRT locus, in cells obtained from animals which have been treated with the test material. Such systems use the principles of mutation detection developed in *in vitro* studies, e.g. resistance of cells to 6-TG treatment, but such mutations are detected in cells from treated animals rather than in cells exposed to chemicals in culture. Naturally the biggest problem with such an *in vivo* system is that the low frequency of such mutations necessitates the examination of a large number of cells. To date such work has been restricted to examination of mutations in cells such as lymphocytes, which can be harvested from animals in relatively large numbers and can be readily cultured and cloned using current tissue culture techniques.

Naturally, to study mutations extensively *in vivo*, it will be necessary to examine cells from those organs and tissues especially sensitive to the action of the test material. It is of course a possibility that methods for the measurement of mutations at the HGPRT locus may not be readily applicable to cells from such tissues. In spite of such problems, if we are ever to abandon the use of tests which rely on the unreliable empirical relationship between *in vitro* mutation and development of tumours in animals and man, it is essential that methods be developed which allow the study of mutation induction in the tissues and organs from animals treated with genotoxic materials.

9.2 CHROMOSOME EFFECTS

The demonstration of a relationship between treatment with carcinogens and chromosomal changes in exposed cells has resulted in the acceptance of cytogenetic damage as an end-point for the detection of genotoxic material. Such studies can rely on the direct recognition and quantitative assessment of clearly recognisable alterations to chromosome morphology or chromatid structure. Other 'indirect' methods of assessing chromosomal damage involve detecting micronuclei, which are chromatin fragments or whole chromosomes resulting from disruption or breakage of chromosomes during cell division. These micronuclei are readily observable, especially in cells with a large cytoplasm and lacking lobed nuclei, and, since they can be seen throughout the whole cell cycle, the special procedures and techniques required for visualising chromosomes in cells are not necessary.

In vitro systems

In vitro metaphase analysis A variety of cell lines has been used in *in vitro* cytogenetic studies, but broadly speaking the most common types are the Chinese hamster ovary cell line or human peripheral blood lymphocytes. The techniques used are relatively simple in that the cultured cells are exposed to the agent under test and then harvested at varying times after treatment. To allow visualisation of the chromosome structures, the cells are arrested in metaphase (usually using a spindle inhibiting agent such as colchicine), swollen using hypotonic solutions, fixed, stained and examined with light microscopy. By careful examination of the mitotic figures various signs of damage, including breaks, gaps, fragments, rings, etc., can be seen. (For a complete description of the various changes that can be detected and the evaluation of resulting data see Scott *et al.* (1983) and Bruisk (1980). The critical factors associated in performing such tests are also described in these references.)

Lymphocytes, which do not normally grow and divide in culture, are stimulated into division by adding the mitogen phytohaemagglutinin (PHA) to the culture medium. As cell division does not begin until about 40 hours after the addition of the PHA, timing of exposure of the lymphocytes to the test materials becomes very important. If the agent is added at the establishment of the culture it may be broken down, spontaneously or by enzyme activity, before the cells begin dividing.

Instead of looking for readily observable changes in chromosome morphology, it is possible to detect and measure reciprocal exchanges between sister chromatids (sister chromatid exchanges; SCEs). Scientists working in genotoxicity research have divided opinions on the value of SCEs in risk assessment. Some suggest that they may be indicative of either chromosome effects or some DNA-damaging event which does not have a mutagenic event as an 'endpoint'. Because the biological significance of SCEs is unknown (e.g. chemicals producing an increase in SCEs do not necessarily produce chromosomal aberrations and vice versa), this technique is generally regarded as a research tool rather than a screening technique. (An outline of techniques used to measure SCEs, together with a discussion of the critical factors which may influence the tests, can be found in the report by the *UKEMS Sub-Committee on Guidelines for Mutagenicity Testing* (Part II). See Scott *et al.* (1983).)

In vivo systems

The use of *in vivo* systems for studying cytogenetic damage naturally has all of the advantages described for *in vivo* mutation assays, e.g. presence of an endogenous metabolising system, comprehensive DNA repair system, etc. Another feature of *in vivo* cytogenetics is that there is, in many instances no need to include a tissue culture phase, which is an essential part of *in vitro* mammalian mutation assays. It is thus possible, theoretically at least, to undertake a direct cytogenetic analysis on any tissue or organ which has a naturally occurring dividing cell population or which can be induced to proliferate (such as the liver, by partial hepatectomy).

Because of the ability to examine gonadal tissue, it could be imagined that it should be possible to study the ability of chemicals to produce chromosomal aberrations in germ cells. In practice, however, the majority of *in vivo* chromosome aberration studies have concentrated on using bone marrow cells or peripheral blood lymphocytes. (In fact the most widely used *in vivo* procedure for measuring chromosomal effects is not by looking directly for chromosomal aberration, but rather by using the more indirect approach of looking for chromosome fragments (micronuclei) in erythrocytes; this is discussed later.) The reason for the general lack of activity in examining solid tissues is the technical difficulty in producing adequate numbers of mitotic figures of good enough quality for examination. Also, when screening new drugs, it is difficult to predict which tissues should be examined to ensure detection of what might be a tissue-specific interaction.

The general approach to *in vivo* cytogenetic analysis has been pragmatic. Assays have been selected which, although by no means perfect, are considered to be helpful in giving some information as to the potential of agents to produce cytogenetic damage. These assays are described below.

In vivo metaphase analysis. The standard procedure is to treat the experimental animals, usually rodents, with the agent under test and, at various times after exposure, e.g. 6, 12, 24 and/or 48 hours, harvest the cells to be examined. There is a variety of techniques which can be followed once dosing has been completed. When performing a bone marrow analysis, the animals can be injected with colchicine, which arrests dividing cells in metaphase, approximately one to three hours before the animals are killed. Bone marrow cells are then harvested and processed for examination. Alternatively, harvested cells can be treated *in vitro* with colchicine and then prepared for metaphase analysis.

Other methods involve examining aberrations in peripheral blood lymphocytes. The technique follows the procedure used for *in vitro* studies, except of course that exposure of the cells to the test material occurs *in vivo*. It may be imagined that the examination of bone marrow metaphase preparations from treated animals, especially those on conventional (e.g. 30 day) toxicity studies, would prove useful in determining the clastogenic potential of novel compounds (i.e. their potential to cause chromosomal damage and/or a change in chromosome numbers). Such an approach has not been widely used, as it has been shown that positive responses seen after a single dose of test agent may disappear after more prolonged treatment. This has been attributed to the fact that repeated dosing may kill 'sensitive' cells, or that the animals may become 'tolerant' of the agent. However, insufficient studies have been conducted to provide information for critical evaluation.

While it is also possible to examine SCEs in cells obtained from treated animals, the problems associated with interpreting the biological significance of SCEs, together with the increased technical difficulties associated with such *in vivo* studies, make it difficult to envisage such studies achieving wide acceptance. However, such difficulties have not prevented a number of experimenters from examining the incidence of SCEs in workers exposed to industrial chemicals. The relevance of such findings to hazard assessment remains unknown.

Micronuclei in bone marrow polychromatic erythrocytes. This is currently the most widely used *in vivo* test system. The assay is based on the measurement of micronuclei remaining in polychromatic erythrocytes following expulsion of the main nuclei after the last mitosis. These 'micronuclei' may be small fragments of chromosomes which have become 'broken off' following exposure to a clastogen, or single chromosomes separated during the formation of the nuclear membrane. Whatever their origin, micronuclei remaining in young erythrocytes can be readily detected as densely staining small bodies in the anucleate cells.

The general procedure adopted for measuring micronuclei is to dose mice with a test agent, by the proposed therapeutic route, up to almost lethal acute doses. At varying times after treatment, usually 24, 48 and 72 hours, groups of animals are killed, the bone marrow aspirated and the polychromatic erythrocytes scored for the presence of micronuclei. The ratio of polychromatic to normochromatic erythrocytes is also usually determined to see whether the test material has any effect on cellular proliferation in the bone marrow.

As stated above, the micronucleus test is perhaps the most widely used *in vivo* assay, even though it has been criticised for being inefficient as well as insensitive. However, because the endpoint is technically easy to measure the assay is routinely performed in many laboratories and, as a result, has become universally accepted by regulatory authorities as a satisfactory test.

Germ cell chromosome analysis. The basis of these studies is to predict transmissible genetic damage, resulting from structural or numerical chromosomal aberrations, to cells of the reproductive tissues. It might therefore be supposed that an assessment of such risk could be best accomplished by examining germ cells, obtained from treated animals, for chromosomal aberrations. Unfortunately it is technically difficult to produce adequate chromosome preparations of either spermatogonial stem cells or resting oocytes, which are the germ cell stages considered to be of greatest use in risk assessment. Furthermore, examination of spermatocytes can be criticised owing to the fact that many cells carrying chromosomal aberrations might be lost following cell division. As a result of these and other problems, cytogenetic analysis of chemical damage in male and female germ cells has been performed on relatively few chemicals. Such studies are therefore only rarely used for assessing the genotoxic potential of compounds.

A method which can be used to measure cytogenetic damage in germ cells is the 'dominant lethal assay' in which genetic changes in parent germ cells, egg or sperm, are detected by measuring deaths in first generation embryos. Dominant lethals reduce litter size by failure of the fertilised egg to implant or develop after implantation. Scoring of dominant lethals usually includes the measurement of all excess deaths, as compared to controls, occurring during embryo development, i.e. lethal effects due to reduced numbers at implantation (pre-implantation loss), or reduced numbers of viable implanted embryos (postimplantation loss). Although a number of methods have been designed for conducting dominant lethal assays, only the most common technique, using treated males, will be briefly described here.

Male mice or rats are exposed to one of three concentrations of test material, the highest dose being at or about the LD_{50}. The animals can be given (a) a single or subacute (five to seven days) treatment followed by a continuous, eight to ten week, mating schedule, or (b) an eight-week continuous treatment (but not at the LD_{50} level) followed by a single mating. The first scheme is the one generally adopted, since weekly mating for eight consecutive weeks is approximately equivalent to the length of the spermatogenic cycle and thus the response of each germ cell stage is observed.

Assessment of any effect is made by counting numbers of live foetuses, and dead foetuses (early deaths are characterised by discrete dark brown areas on the endometrium representing a necrotic mass of placenta). A variety of statistical methods can be used to interpret the data in terms of positive and negative responses.

The fact that the dominant lethal test is a relatively simple assay to perform, can be done quickly and the endpoint is not subjective (i.e. embryos are either dead or alive), has made this procedure the most widely used for detecting cytogenetic changes in germ cells. The test does not, however, evaluate genetic damage transmissible to the progeny and is considered by many to be relatively insensitive.

The sperm morphology assay has also been advocated as being able to detect potential germ cell mutagens. It is technically simple to perform and is cheap and rapid. The assay consists of treating male mice or rats, sacrificing them at weekly intervals and examining the epididymal fluid. The genetic basis for the production of sperm with morphological abnormalities is, however, in dispute. Also, it is thought that most of the abnormal sperm are reabsorbed.

Although, as shown above, there are several assays designed to detect chemicals which cause genetic damage to germ cells, in practice such tests are rarely used. The reason for this is that the germ cell mutation assays at present available appear to be intrinsically insensitive in detecting chemical mutagens. This insensitivity was emphasised in a study by Holden (1982), who reported that whereas known germ cell mutagens could be adequately detected using somatic cell mutation assays, a number of these agents was inactive in germ cell tests. In spite of these findings, some regulatory authorities appear to be greatly concerned with the theoretical prospect that certain chemicals may be germ cell mutagens without producing effects in somatic cell tests. At this time, for screening purposes the obligatory use of germ cell mutation tests does not appear to be justified.

9.3 DNA INTERACTIONS

This category of tests is based on the theory that chemicals capable of interacting with DNA have a high probability of being carcinogenic. Thus various assays have been developed which do not measure mutations or chromosomal damage *per se*, but rather measure direct binding of chemicals to DNA, physical damage to DNA caused by such interactions (e.g. strand breaks), or cellular repair of such damaged sites (e.g. unscheduled DNA synthesis).

DNA binding

There are several methods available to determine whether or not a chemical will bind directly to DNA-forming chemical adducts. Such methods may involve exposing purified DNA, and/or cells in culture, to the test agent, or treating experimental animals and then analysing the tissue of interest. Most of such studies use radiolabelled materials of high specific activity and such isotopically labelled materials may not be available for experimental drugs requiring screening.

An alternative approach to looking for adducts has been to measure the damage which such interactions produce in DNA. Several methods have been reported as being useful in identifying such damaged sites. These include measurement of the reduction in the size of single stranded DNA using caesium chloride gradients or the sensitivity of damaged sites to treatment with alkaline solutions. Although these tests are relatively simple and rapid to perform, they have not been extensively validated by comparing related carcinogenic and non-carcinogenic materials.

DNA repair – unscheduled DNA synthesis (UDS)

The most commonly used method of determining the ability of a chemical to interact with DNA is not to measure DNA damage directly, but to demonstrate the repair of such damage. In this assay non-dividing cells, e.g. cells in which DNA synthesis has been blocked or cells which do not normally divide in culture (e.g. hepatocytes), are exposed to the test agent, followed by incubation in a medium containing ^{3}H-thymidine. Cells undergoing DNA repair incorporate the radio-isotope into their DNA, which is then measured by autoradiography or scintillation counting techniques.

It is now possible to perform this assay using experimental animals rather than cultured cells. For the *in vivo* study, animals are exposed to the agent under investigation for varying times, then killed and the tissue(s) of interest, e.g. liver, gonads, etc. are removed. Individual cells are prepared, usually by enzymatic disaggregation and incubated in a culture medium containing ^{3}H-thymidine. Any repair synthesis is once again detected by autoradiography or by scintillation spectrometry of DNA isolated from the cell lysates.

9.4 NEOPLASTIC CELL TRANSFORMATION

Over the years there have been a number of attempts to develop *in vitro* transformation assays as a relevant endpoint to detect carcinogens, irrespective of their mechanism of action. Transformation of cells *in vitro*, as *in vivo*, is usually detected by phenotypic changes and it has been shown that certain morphological changes which can be seen in culture are associated with development of tumours when such 'morphologically transformed cells' are inoculated into appropriate recipient animals. As a result of these observations, several tests have been published which claim to be capable of detecting the oncogenic potential of chemicals by the induction of such 'neoplastic transformations' in mammalian cells in culture.

One of the first such assays which was widely advocated as being useful for detecting carcinogens, was the BHK21 cell transformation assay developed by Styles (1977). This assay relied on the development of transformed foci, which were detected by their ability to grow and form colonies in soft agar rather than by any readily observable change in morphology. After much interest, this test is now little used owing to difficulties in reproducing consistent results in different laboratories.

The more common endpoint used at present is morphological transformation in fibroblast cultures, such as the C3H10T1/2 and BALB 3T3 cell lines. The general procedure is to expose low numbers of cells to the agent under test and then allow the cultures to continue growing for at least six weeks. The transformed cells are characterised by the piling up of cells in a criss-cross pattern, representing a loss of contact inhibition. The non-transformed cells retain their morphology and form a monolayer.

9.5 REGULATORY REQUIREMENTS FOR GENOTOXICITY TESTING

Genotoxicity testing of new drugs is now a regulatory requirement of many countries. Because of the international nature of the major pharmaceutical companies, in both marketing and conducting clinical trials, it is essential that they take account of the demands of genotoxicity testing in all countries. However, because genotoxicity testing requirements vary from country to country and test systems are constantly evolving, it is impossible to be precise about the exact tests which may be required. In general, however, the EEC requirements for mutagenicity testing are the most stringent, and companies satisfying these requirements will, in most instances, cover the demands of other countries. For this reason only the regulatory requirements of the EEC will be described in detail here, with specific requirements of other countries being briefly mentioned.

Before new drugs can be marketed in the EEC, they must be tested for genotoxic potential. To fulfil such a requirement it is necessary to perform appropriate assays from a 'base set' of tests. This base set is composed of four categories (see below) and it is usual to perform a single assay from each category. The choice of assay is the responsibility of the investigator.

Base set of tests
  (i) gene mutation in bacteria
 (ii) *in vitro* test for gene mutation in mammalian cells
(iii) *in vitro* test for chromosome aberration in mammalian cells
(iv) *in vivo* test for genetic toxicity

Within this basic framework, member countries of the EEC retain the right to demand other tests. It is hoped, however, that a dossier containing appropriate assays from each of the base set of tests will be generally acceptable not only to EEC member states, but also to other countries worldwide.

While such tests are necessary for registration of drugs, certain countries, e.g. the UK and Australia, also demand some mutagenicity testing, e.g. an Ames test, prior to clinical trials approval. Other countries, such as the USA, currently have no requirement for mutagenicity testing of new drugs prior to either clinical trials or registration. However, if tests have been carried out for other countries their results must be submitted to the USA authorities.

Another important consideration is, of course, how the data from such studies are received and interpreted by the authorities. The feeling is that some countries take a pragmatic attitude, and believe that results of such tests may not provide information that is helpful in the assessment of the risk—benefit considerations for a new drug. Other authorities may reject a product registration application simply on the basis of a positive result in one of the bacterial strains used in the Ames test.

9.6 DISCUSSION

In the preceding pages we have seen something of the routine assays used in genetic toxicology and how the requirement for genotoxicity testing has been adopted by regulatory authorities. Indeed provision of genetic toxicity data is now a necessity for the worldwide development and registration of new drugs.

Theoretically, mutagenicity testing is based on an impressive background of basic scientific knowledge and information on genetic effects. Because of this, it is assumed that knowledge of such experimental systems would provide exploitable assays capable of detecting potentially genotoxic agents. In practice, however, the predictive nature of such tests has been found to be less than reliable, and shortcomings have been demonstrated in a number of studies undertaken to evaluate their performance. For example, Shelby & Stasiewicz (1984) reported that 60% of a group of chemicals found to be non-carcinogenic in experimental animals showed positive effects in *in vitro* genotoxicity studies, while Zeiger & Tennant (1986) reported that only about 50% of chemicals which had produced tumours were active in *in vitro* tests. Such data have resulted in a number of investigators questioning the ability of genotoxicity tests to identify carcinogens accurately.

It is obvious that quick, efficient assays to identify potentially mutagenic and carcinogenic compounds would be most useful to the pharmaceutical industry, allowing the elimination of such materials before initiating extensive and expensive developmental programmes. On the other hand, if such assays generate false positive results, this could lead to the loss of valuable pharmaceuticals. Further consequences of this may include prolonged suffering of individuals, who could gain benefit from treatment and loss of revenue for the company, which could have been used for the development of other much needed drugs.

The obvious need is to eliminate genotoxicity tests which perform badly (cynics would say all genotoxicity tests perform badly). This is not easy to do, since each particular test has its devoted advocates, and because of the legion of studies carried out the 'sensitivity' and 'specificity' of tests may swing one way or another depending where, when, how and by whom such studies are performed. It has been suggested that a multiplicity of tests should reduce areas of uncertainty. A natural problem of course is that the imperfections of these assays may be magnified when combined into a battery of tests. Because of the influence such results can have on regulatory authorities, positive results could well cause the demise of a new compound. Therefore what is required is a rational scientific approach to the interpretation of data.

In spite of the fact that short term genotoxicity assays suffer from problems (some more than others) of reproducibility, sensitivity and predictive value, such tests are now used by most if not all pharmaceutical companies. Naturally no company wants to expose man to dangerous chemicals, or indeed inadvertently alter the genetic endowment of the human species. It is, however, necessary to see such tests as part of the overall safety evaluation, i.e. in combination with data on animal toxicity, pharmacokinetics, pharmacology, etc. The key word must be 'overall' and too much emphasis must not be given to positive genotoxicity data. This situation may, of course, be different if there is evidence early in development that the compound is producing suspicious lesions or is a potential alkylating agent, etc.

While we look at positive genotoxicity results with some suspicion, it is important not to get a false feeling of comfort from negative responses. It is quite surprising the number of *in vivo* genotoxicity assays which are undertaken with species not normally used in toxicological assessments, sometimes without any evidence that the drug is even absorbed, let alone gets to the tissues of interest. Also, in many *in vitro* assays, the test chemical may be dissolved in some organic solvent without any evidence of stability, whether it precipitates on addition to the culture medium, or even if it gets into the cells.

In conclusion we must approach genotoxicity testing with caution, attempting to get a correct balance between scientific judgement and regulatory requirements.

REFERENCES

Ames, B. N., McCann, J. & Yamasaki, E. (1975) Methods for detecting carcinogens and mutagens with the salmonella/mammalian microsome mutagenicity test. *Mutation Research*, **31**, 347–64.
Bruisk, D. (1980) *Principles of Genetic Toxicology*. Plenum Press, New York.

Farrow, M. G., McCarroll, N. E. & Anletta, A. E. (1986) 1984 survey of genetic toxicology testing in industry, government and academic laboratories. *Journal of Applied Toxicology*, 6, 211–23.

Scott, D., Dandford, N., Dean, B., Kirkland, D. & Richardson, C. (1983) *In vitro* chromosome aberration tests. In *UKEMS Sub-Committee on Guidelines for Mutagenicity Testing*. United Kingdom Environmental Mutagen Society.

Shelby, M. D. & Stasiewicz, Z. (1984) Chemicals showing no evidence of carcinogenicity in long term two species rodent studies. The need for short-term test data. *Environmental Mutagenicity*, 6, 817–78.

Styles, J. A. (1977) A method for detecting carcinogenic organic chemicals using mammalian cells in culture. *British Journal of Cancer*, 36, 558–64.

Zeiger, E. & Tennant, R. W. (1986) Mutagenesis, clastogenesis, carcinogenesis, expectations, correlations and relations. *Progress in Clinical Biological Research*, 209, 75–84.

APPENDIX 1

Toxicology data requirements for clinical trial approval and registration of new drugs

This section has been included to provide a summary of the types of toxicology data required by the major regulatory authorities. It must be emphasised that it is only a summary and anyone preparing a registration dossier should refer to the original toxicity guidelines.

(Addresses of the regulatory authorities are given in Appendix 2, p.181–182.)

1. SINGLE ADMINISTRATION (ACUTE) TOXICITY STUDY

Australia

Animal species: At least two, one usually being a non-rodent other than a rabbit. (This requirement is not normally implemented and studies in two rodent species, e.g. rats and mice, are normally acceptable. However, if large differences in response are seen, then a third species should be included.)

Number of animals: In principle, the same numbers of male and female animals should be used. No exact number is given and a formal estimate of the LD_{50} is not required.

Administration route: In general, two routes of administration should be used. The proposed therapeutic route must be used in both species, with the oral route or injection in at least one of them.

Observations and examinations: Animals should be observed for not less than 10 days, usually 14, and adequate examinations should be undertaken to reveal tissue or organ damage.

Canada

Animal species: At least three species must be used (two rodent and one non-rodent other than the rabbit). Additional species may be required if large differences in response are seen.

Number of animals: For rodents, ten of each sex per dose are required to determine the LD_{50} (with confidence limits), the minimum lethal dose (MLD) or a tolerance level. With non-rodents, smaller numbers can be used, but they must be sufficient to provide an approximate LD_{50} or MLD.

Administration route: The proposed clinical route plus intravenous (IV) dosing should be used.

Observations and examinations: Animals should be observed for at least ten days and the onset, duration and severity of symptoms noted. Time and cause of death should be recorded and, in the event of delayed deaths, an autopsy, sometimes including histopathology, may be of value.

EEC

Animal species: At least two mammalian species of known strain should be used.

Numbers of animals: Where possible, the LD_{50} value with its confidence limits should be determined. Equal numbers of male and female animals must be used.

Administration route: Two different routes should be used, one being identical or similar to the proposed route in humans and the other ensuring systemic absorption of the substance.

Observations and examinations: The period during which the animals are observed should not be less than a week.

Japan

Animal species: At least two species must be used, one usually being a non-rodent other than the rabbit.

Numbers of animals: At least five animals per group are required in rodents, two per group in non-rodents (males and females should be used in at least one species). The numbers should be adequate to provide an approximate LD_{50} in rodents and clear toxic signs in the non-rodent.

Administration route: In principle, oral and parenteral routes should be used, including the expected clinical route.

Observations and examinations: Animals should usually be observed for 14 days during which time type, grade, time of appearance, progression and reversibility of toxic signs should be recorded. Autopsy of all animals and histopathology of any abnormal tissues are required.

Nordic

Animal species: At least two suitable species must be used.

Numbers of animals: Adequate numbers of rodents are required to provide a quantitative evaluation of the LD_{50}, although a high degree of precision is not required. In addition to the LD_{50} in rodents, an estimate of the highest tolerated dose and/or lowest lethal dose in non-rodents may be required.

Administration route: Various methods of administration should be used.

Observations and examinations: Animals should be observed for seven days or longer. Toxic effects and times of their appearance should be noted, together with the mode of death. (Acute toxicity in newborn animals is required for drugs that may be used during the perinatal period.)

UK

Animal species: At least two suitable species must be used.

Numbers of animals: Equal numbers of males and females should be used. However, if no sex differences are seen in the first species, then only one sex may be used in subsequent studies. A quantitative evaluation of the LD_{50} and information on dose–response relationships, without a high degree of precision, are required.

Adminstration route: At least two routes of administration are necessary, one identical with, or similar to, the proposed therapeutic route, the other ensuring systemic exposure. If the proposed route of administration in man is IV, then this route alone is acceptable.

Observations and examinations: Animals should be observed for at least seven days and any signs of toxicity and mode of death determined. Histopathological examination of any abnormal tissues seen at necropsy should be undertaken.

USA

Animal species: At least three mammalian species are required, one must be a non-rodent.

Numbers of animals: Sufficient numbers of rodents should be used to provide an LD_{50} with confidence limits. For non-rodents, an approximate determination will suffice.

Administration route: The LD_{50} should be determined by the proposed human route and one other route.

Observations and examinations: Animals should be observed for at least one week, longer if there are still overt signs of toxicity and delayed deaths occur.

General comments

In the classic LD_{50} test, it was necessary to use sufficient numbers of animals and dose groups to determine the dose that would kill 50% of the animals with reasonable statistical accuracy. It is now becoming evident that a precise estimation of the LD_{50} is, with the exception of a few highly toxic drugs such as cancer chemotherapeutic agents, unnecessary . Other measurements, such as on sites and mechanisms of action, provide a better measure of the hazards involved.

2. REPEAT ADMINISTRATION TOXICITY STUDIES
Australia

Animal species: At least two species must be used, one of them being a non-rodent other than the rabbit.

Numbers of animals: No exact numbers are given.

Administration route and duration of exposure: Treatment must be given by the proposed therapeutic route in man. When the drug is given orally, evidence of absorption should be provided, and when it is incorporated into food or water, drug stability and intake should be monitored. The duration of treatment depends upon the proposed duration of treatment in humans:.

>   Single or several doses on one day in humans
>> – 2 weeks in animals
>
>   A few days in humans
>> – 4 weeks in animals
>
>   Up to one month in humans
>> – 12 weeks in animals
>
>   Over one month in humans
>> – 26 weeks, or longer, in animals

Observations and examinations: Although it is stated that the highest dose must produce toxicity, no indications are given as to the types of measurements which should be made.

Canada

Animal species: At least two species must be used, one of them being a non-rodent other than the rabbit.

Numbers of animals: Sufficient numbers of rodents should be used to permit periodic laboratory investigations, histopathology and information on reversibility of toxic effects. With non-rodents, small numbers can be used.

Administration route and duration of exposure: If a drug is to be given to man daily, it should be administered to experimental animals seven days a week by the proposed therapeutic route. When the drug is given orally, evidence of absorption is required, together with correlations between animal and human blood levels of the parent drug and the principal metabolite(s). The duration of treatment in experimental animals depends upon the proposed duration of exposure in humans:

Up to a week in humans –

4 to 6 months in animals

Longer than a week in humans –

18 months in animals

Observations and examinations: Compound-related changes in behaviour, appearance, neurological effects, food and water consumption, body weight, haematology, clinical chemistry and urinary parameters should be recorded. Eye examination and cardiovascular, as well as physiological and pharmacological, manifestations should be monitored. At necropsy, major organs should be weighed and examined for gross pathological changes. Histopathological examinations must be conducted on control and high dose animals, with tissues from lower dose animals being included if gross or microscopic changes were seen at the higher doses.

EEC

Animal species: At least two species should be used, one a non-rodent, chosen on the basis of their similarity to man.

Numbers of animals: No exact numbers are given, although consideration should be given to having adequate numbers to determine all relevant toxicological effects.

Administration route and duration of exposure: Administration should be via the route intended for use in man with treatment seven days per week. However, if elimination of the drug is very slow or fast, it may be necessary to dose less than or more than once per day respectively. Duration of treatment in experimental animals depends upon the proposed duration of exposure in humans:

Single or several doses in one day in humans
    – 2 weeks in animals
Up to 1 week in humans   – 4 weeks in animals
Up to 30 days in humans   – 3 months in animals
More than 30 days in humans – 6 months in animals

Observations and examinations: Compound-related changes should be recorded food intake, body weight, haematology, clinical chemistry, urinalysis, ophthalmoscopy, general behaviour and ECG. All animals must be subjected to necropsy, with the following tissues taken for histology: gross lesions, tissue masses or tumours (including regional lymph nodes), blood smears, lymph nodes, mammary glands, salivary glands, sternebrae, femur or vertebrae (including bone marrow), pituitary, thymus, trachea, lungs, heart, thyroid, oesophagus, stomach, small intestine, colon, liver, gall bladder, pancreas, spleen, kidneys, adrenal glands, bladder, prostate, testes, ovaries, uterus, brain, eyes, spinal cord. In rodents all tissues from the control and high dose animals must be examined, with examination of the lower dose groups being restricted to those tissues showing pathological changes. In other species, where small numbers of animals are used, histopathology should be conducted on all animals.

Japan

Animal species: Two species must be used, one of them selected from among non-rodents other than rabbits.

Numbers of animals: Studies lasting 28 to 90 days should include at least ten male and ten female animals per group for rodents and at least three males and three female non-rodents. For longer studies more animals should be used, e.g. for 6 to 12 month studies at least twenty rodents of each sex per group and for non-rodents at least four of each sex per treatment group.

Administration and duration of exposure: In principle the expected clinical administration route should be used, with treatment being performed seven days a week. The period of clinical use of the test substance is based on the following standards:

Single administration or repetitive
  administration for up to 1 week in humans
                                        – 1 month in animals
Up to 4 weeks in humans   – 3 months in animals
Up to 6 months in humans   – 6 months in animals
Long term repetitive administration in humans
                                        – 12 months in animals

Observations and examinations: The frequency with which observations and examinations are carried out should be scheduled by considering the animal species, the type and manifestation period of toxic signs, and the effect of these operations on the animals. Observations and examinations include body weight, food intake, water intake, haematological examinations, clinical chemistry, urinalysis, ophthalmoscopy and, if necessary, ECG and visual, auditory and renal function tests should be performed.

All animals should be necropsied, with the following tissues being selected for histopathological examination: heart, liver, spleen, kidneys, adrenal glands, prostate, testes, ovaries, brain, hypophysis (these are virtually always weighed prior to fixation), lungs, salivary glands, thymus, thyroid gland, seminal vesicles, uterus (often weighed), mammary gland, salivary gland, bone and bone marrow (sternum, femur), trachea, bronchii, stomach, parathyroid glands, tongue, oesophagus, duodenum, small intestine, large intestine, gall bladder, pancreas, urinary bladder, vagina, spinal cord, eyeballs, lacrimal glands, harderian glands and any other sites with macroscopic changes.

Nordic

Animal species: At least two mammalian species must be used, one should be a non-rodent. The species should be selected on the basis of pharmacokinetics, pharmacodynamics, previous toxicity testing and comparison to human pharmacology.

Numbers of animals: No exact numbers are given.

Adminstration and duration of exposure: Drugs intended for local application should provide information about the drug's direct effect on the relevant tissue. Special attention should be paid to data on absorption. Duration of treatment in experimental animals depends upon the proposed duration of treatment in humans:

Exposure in man one to seven days – minimum two to four weeks in experimental animals (two species)

Exposure in man up to 30 days – minimum of three months toxicity studies in experimental animals (two species)

Exposure in man longer than 30 days – minimum of six months toxicity data in experimental animals (two species).

Observations and examinations: Study of the reversibility of toxic effects may be useful. Examinations include weight gain, food and fluid intake, clinical observations, mortalities, haematology, clinical chemistry, blood drug concentrations, some functional examinations e.g. ECG, ophthalmoscopy, blood coagulation, urinalysis, macroscopic examination at necropsy, organ weights, histopathological examinations of organs and tissues with special emphasis on suspected target organs.

UK

Animal species: At least two mammalian species must be used, one a non-rodent. The selected species should metabolise the drug in a manner as similar to man as possible. The drug should be able to demonstrate pharmacological activity in the species and strain used. If the compound under study demonstrates melanin binding, then albino animals should not be used in the toxicity studies.

Numbers of animals: No numbers are given, but treatment groups should be large enough to reveal all toxicologically important effects. The numbers should also be large enough to allow interim kills and some animals to be retained for reversibility studies.

Administration and duration of exposure: Administration should be via the proposed route in man, the intraperitoneal route is not recommended unless proposed as the clinical route. Treatment should be given seven days a week; if rate of elimination of the drug is slow, then dosing may be less frequent, but if elimination is rapid, then dosing more than once a day is recommended. Pharmacokinetics studies should be used to measure the quantity of drug absorbed from the proposed site of administration. Duration of usage in man is based upon the following standards:

> Single or several doses in one day in humans
> > – 2 weeks in animals
> Repeated dose up to seven days in humans
> > – 4 weeks in animals
> Repeated dose up to 30 days in humans
> > – 3 months in animals
> Repeated dose for longer than 30 days in humans
> > – 6 months in animals

Observations and examinations: Food consumption, body weight, general behaviour, general health, biochemistry, haematology, urinalysis and (preferably) ophthalmoscopy should be recorded. Necropsies must be performed on all animals, as must histopathology of all organs (see below) from high dose and control group animals, and of any tissues from any animals in any groups in which macroscopic lesions are found at necropsy. Tissues to be taken for histological examination include: gross lesions, tissue masses or tumours, blood smears, lymph nodes, mammary glands, salivary glands, sternebrae, femur or vertebrae, pituitary, thymus, trachea, lung, heart, thyroid, oesophagus, stomach, small intestine, colon, liver, gall bladder, pancreas, spleen, kidneys, adrenal glands, bladder, prostate, testes, ovaries, uterus, brain, eyes, spinal cord. (All of the above listed tissues should be examined from all animals when only four of five animals per group are used.)

Some mention is made of immunotoxicity and if any of the examinations suggest an effect on the immune system, then this should be examined using tests appropriate to the current state of knowledge.

USA

Animal species: Two species should be used, usually the rat and dog. Occasionally a third species, e.g. the monkey, is also used.

Numbers of animals: Numbers should be sufficient to provide a valid estimation of the incidence and frequency of toxic effects. Ten to 25 rodents should be used per sex per group; two to three non-rodents per sex per group. The group size will depend upon toxicity and/or mortality findings from the acute studies, inclusion of interim sacrifices, reversibility evaluation and assessment of other aspects of toxicology.

Administration and duration of exposure: Daily single doses must be given seven days a week using the route of administration proposed for man. In rodents the drug may be administered in the diet if it is not practical to dose by gavage. The recommended duration of animal studies to support the three phases of clinical testing are as follows:

Limit of human drug usage	Phase of clinical study	Duration of animal toxicity studies
1 to 3 days	I, II, III, NDA	2 weeks
Up to 4 weeks	I, II	4 weeks
	III, NDA	13 weeks
Up to 3 months	I, II	13 weeks
	III, NDA	26 weeks
3 months or longer	I, II	13 weeks
	III,	26 weeks
	NDA	52 weeks or longer

(NDA = New Drug Application)

Observations and examinations: The following are required: clinical observations, clinical chemistry, haematology, ophthalmological examination, organ function tests, biopsy, necropsy, histopathology and alteration of cellular structures by electron microscopy and other special techniques. At necropsy routine weighing of at least the liver, kidney and endocrine glands must be carried out. Histopathological examination should include all tissues of at least ten animals per group from the high dose and controls, and from lower dose groups if drug-related lesions are found in the high dose animals. All tissue from all non-rodents must be examined.

General comments

The length of toxicity studies required for drug registration and clinical trials varies among different authorities. In Europe 6 month studies are the longest, while in the USA and Canada 12 and 18 month studies in two species are required for registration. At present there is some discussion as to whether or not there is evidence for toxicological effects that show up between 6 and 18 months of treatment, which are not seen in shorter treatments, and that are not evident in carcinogenicity studies.

3. ONCOGENICITY STUDIES

Australia

Animal species: Two species must be used, with selection being made on the basis of metabolic and pharmacokinetic similarity to humans. Species and strains susceptible to a particular carcinogen or class of carcinogens must be considered. Species and strains with a high or variable incidence of spontaneous tumour formation should be avoided.

Numbers of animals: Fifty animals of each sex should be used in each group.

Administration route and duration of exposure: Administration is normally daily by the route proposed for use in man. Rats require at least 24 months' treatment; mice and hamsters at least 18 months'. Where survival rate is high, there may be advantages in extending these times. Treatment in the drinking water or in the diet is acceptable, but stability data are required. Parenteral administration on a frequent basis may cause local pathology, predisposing the area to neoplastic transformation and thereby causing tumours unrelated to the intrinsic carcinogenicity of the drug. In this case weekly or less frequent dosing may be considered.

Observations and examinations: If signs of toxicity appear during the course of the study, appropriate action should be taken to ensure that adequate numbers of animals survive to the end of the study.

Canada

Animal species: Two species must be used, usually the rat, mouse or hamster. In special cases dogs, primates or other species may be used. Metabolic pathways in man and in the animal species must be considered. Spontaneous tumour incidence, sensitivity to tumour induction, availability, genetic stability, etc., of the species and strain selected should be known.

Numbers of animals: At least 50 animals per sex per treatment should be used. Two control groups each containing 50 animals per sex are recommended.

Administration route and duration of exposure: Administration should be the same as that proposed for use in man.

When using the oral route, dosing may be by gavage, in the diet or in drinking water (stability studies are necessary). With parenteral routes of administration, if treatment is associated with local pathology which may give rise to tumours by mechanisms unrelated to the carcinogenicity *per se* of drug, then consideration must be given to the possibility of less frequent dosing (usually daily). Duration of exposure should cover the greater part of the animals' life span.

Observations and examinations: General care and environmental conditions under which animals are housed must be kept under control. Diets free from common carcinogens should be used (semisynthetic diets should be considered). Animals should be checked frequently, once in the morning and once in the afternoon, seven days a week. The time of onset, location, size and growth characteristics of unusual tissue masses should be recorded. Autopsy must be performed on each animal, with histopathology on all organs and tissues of high dose and control group animals. Organs and tissues showing gross changes in animals of other groups should be examined, as well as organs showing increased tumour incidence in high dose groups.

EEC

Animal species: Two species in which the metabolic handling of the drug is known should be used, preferably species showing similarities to man. Species and strains with high incidence of spontaneous tumours should normally be avoided, whereas those known to be sensitive to one or more carcinogens should be selected.

Numbers of animals: For routine studies with rats, mice and hamsters it is suggested that 50 animals per sex are used for each treatment group, with two control groups each dosed with the vehicle by the same route.

Administration route and duration of exposure: Where possible, dosing should be conducted by the proposed clinical route of administration; where relevant, evidence of absorption should be provided. The frequency of dosing will normally be daily, with treatments lasting 24 months in rats and 18 months in mice and hamsters. If survival is high, there may be an advantage in extending the studies to 30 months in rats and 24 months in mice or hamsters, or for the lifespan of the animals, i.e. to 20% survival in the controls.

Observations and examinations: It is suggested that haematological and biochemical investigations may be helpful in the interpretation of any lesions which are found. Particular attention should be paid to the site of application, if the drug is administered other than by the oral route. A full necropsy should be made on all animals dying during the study or killed because of their poor clinical condition. At the conclusion of the study all surviving animals are killed and a full necropsy conducted on each animal. Histopathological examination of all tissues and organs from high dose and control animals should be undertaken, together with tissues from any animal from any treatment group in which a macroscopic lesion is seen. If a high incidence of tumours is seen in one or more tissues of the high dose group, then these tissues should be examined in the medium and low dose groups even when they appear macroscopically to be normal.

Japan

Animal species: Both male and female animals from two species should be selected. It is desirable to use animals that have shown normal growth and are of the same age (up to the age of six weeks at the start of the trial).

Numbers of animals: Each group should comprise no more than 50 males and 50 females. Allocation of the animals to each group should be made using a proper random sampling method.

Adminstration route and duration of exposure: In principle, the expected clinical application route should be used. Treatment should last from 24 to 30 months in rats and from 18 to 24 months in mice or hamsters, with administration performed seven days a week. Studies should be terminated at the end of treatment, or 1 to 3 months after the end of administration. The maximum experimental period should be 30 months for rats and 24 months for mice or hamsters. When cumulative mortality reaches 75% in the lowest treatment or control group, the survivors should be killed and the study terminated.

Observations and examinations: All animals should be observed daily for general condition. Body weight should be recorded once a week for three months after the start of treatment and thereafter once a month. Animals which die or are killed *in extremis* during the course of the study should be necropsied immediately, and macroscopic and histopathological examinations of organs and tissues performed. At the end of the study all survivors should be subjected to necropsy. Histopathological examination should be performed on all animals in the high dose and control groups. However, organs and tissues found to differ in the incidence of neoplastic lesions between the high dose group and controls will be examined histopathologically in the medium and low dose animals. It is suggested that, at the time of necropsy, it may be desirable to take blood samples to make blood smears and measure peripheral red cell and white cell counts. These measurements may be useful in cases suggestive of blood disorders such as anaemia and swelling of lymph nodes, liver or spleen.

Nordic

Animal species: Usually two species should be used. (There are no other recommendations although countries such as Sweden do have some requirements, i.e. at least 20 animals of each sex should survive to the end of treatment, the animals should be treated over the greater part of their life span: mice – 18 months, rats – 24 months, histological examination should be carried out on all macroscopic lesions and in tissues where tumours may be easily overlooked.)

UK

Animal species: Two species should be selected on the basis of metabolic similarity to man. No species with a high or variable incidence of spontaneous tumour formation should be used. Rats (outbred strains) and mice (F1 hybrids) or hamsters (outbred), with low incidence of neoplasms and high longevity, are recommended.

Numbers of animals: At least 50 animals per sex per group should be used; two control groups are recommended.

Administration route and duration of exposure: Administration should be the same as proposed for clinical use where possible. Evidence of absorption should be provided. Treatment should be daily, lasting for 24 months in the rat and

18 months in the mouse or hamster. Where malignancy of tumours is in doubt, treatment may be extended to 30 months in rats and 24 months in mice and hamsters.

Observations and examinations: Animals should be regularly observed for general health and examined clinically to determine incidence and time of appearance of tissue masses. Food consumption and body weight should be recorded at regular intervals. Full necropsy should be performed on all animals, with microscopic examination of all tissues and organs from high dose and control animals, as well as from medium and low dose groups when necessary.

USA

Animal species: Two species should be used. The rat and mouse are usually most appropriate.

Numbers of animals: At least 50 animals per sex per group must be used.

Administration route and duration of exposure: The proposed human route should be used, with daily treatments for at least 24 months in the rat and at least 18 months in the mouse; otherwise the study should be terminated when mortality reduces the control group to 40% of its original size.

Observations and examinations: Complete necropsy of all animals must be conducted. Histopathological examination of all tissues from high dose and control groups is required. If there are questionable findings in the high dose animals, then appropriate tissues from the medium and low dose animals should be examined.

4. REPRODUCTION STUDIES

Australia, Canada, EEC, Nordic countries, UK, USA

	Segment I	Segment II	Segment III
Animal species: *Animal numbers:*	One, usually rat. At least 24 females and 24 males.	Two, usually rat and rabbit. At least 20 rodents. At least 12 non-rodents.	One, usually rat. At least 12.
Treatment:	Animals dosed daily over period of gamete maturation (Males 70 days prior to mating, females 14 days prior to mating). Treated males may be mated with treated females. Dosing in females continued until weaning.	Pregnant females dosed daily throughout period of organogenesis. Foetuses obtained by Caesarean section shortly before time of parturition.	Dosing of pregnant females throughout last $\frac{1}{3}$ gestation until weaning. Animals allowed to litter naturally.
Terminal studies:	At least 10 females killed before parturition, uterus examined for live, dead and malformed embryos. Remainder allowed to litter normally. Young may be observed up to weaning or sexual maturity.	Uterus of mother examined for total number of implantations and resorptions, the placentae for anomalies and ovaries for number of *corpora lutea*. Foetuses should be examined for external malformations, skeletal and visceral abnormalities.	The following assessments should be made: labour, delivery, numbers of live and dead young, malformations, litter size, lactation, neonatal viability.

Japan

	Segment I	Segment II	Segment III
Animal species:	One, usually rat.	Two, usually rat and rabbit.	One, usually rat.
Animal numbers:	At least 20 per sex per group.	At least 30 rodents. At least 12 non-rodents.	At least 20.
Treatment:	Males dosed for 60 days or more before mating, females at least 14 days before mating. After successful mating the females dosed for further 7 days.	Pregnant females dosed daily throughout period of organogenesis (rats – days 7 to 17, mice – 6 to 15 and rabbits – 6 to 18).	Pregnant females dosed continuously from end of organogenetic period until weaning (mouse – day 15 to 21 after delivery, rat – day 17 to 21 after delivery).
Terminal studies:	All females killed just prior to parturition. Numbers of *corpora lutea*, successful pregnancy, dead foetuses, etc. recorded. Gross examination of organs and tissues of dams. Necropsies of males and females where mating unsuccessful.	All rabits and ⅔ rodents killed just prior to parturition. Remainder of rodents allowed to deliver and nurse their young. Foetuses examined for external malformations, skeletal and visceral abnormalities.	Examination of dams for mortality, delivery of young, nursing, etc.

5. GENOTOXICITY STUDIES

Australia

Several tests, selection of which should be justified by the investigator, must be carried out on all drugs which are:

(i) to be used over periods of years (particularly in children),
(ii) chemically related to known carcinogens,
(iii) known to depress bone marrow at tolerated dose levels,
(iv) teratogenic or reduce fertility in reproductive studies.

Canada

No battery of tests is recommended, but assays should detect point mutation, chromosomal aberrations and sister chromatid exchanges. Such studies are required for the same classes of drugs as those listed for Australia (above).

EEC

Requirements are for tests (*in vitro* and *in vivo*), to identify chemicals with mutagenic properties, with maximum accuracy at reasonable cost. Recommended methods include:

(i) tests for gene mutation in bacteria,
(ii) tests for *in vitro* chromosomal aberrations,
(iii) gene mutation tests in mammalian cells,
(iv) *in vivo* tests for genetic damage.

Japan

All new drugs should be subjected to mutagenicity studies with gene mutation and chromosomal aberrations as indices. Recommended assays include:

(i) bacterial mutation,
(ii) *in vitro* chromosomal aberrations,
(iii) *in vivo* micronucleus tests.

Nordic

A small number of well-established *in vitro* and *in vivo* tests, covering both gene mutation and chromosome damage, should be carried out.

UK

As for EEC.

USA

At present the FDA does not require mutagenicity testing for human drugs, but usually recommends such studies for particular categories of drugs such as anti-virals, anti-psoriatics and immunosuppressants. If genotoxicity studies have been undertaken, they must be submitted and if microbial tests show positive or questionable results, it is recommended that mammalian cell assays are used.

APPENDIX 2

Countries and addresses of regulatory agencies

Australia: Commonwealth Department of Community Services and Health, PO Box 100, Woden ACT, 2606 Australia.

Canada Health and Welfare Canada Tunney's Pasture, Ottawa, Ontario, Canada K1A OL2.

EEC: Directorate-General for Internal Market and Industrial Affairs, Rue de la Loi 200, B-1049 Brussels, Belgium.

Japan: Pharmaceutical Affairs Bureau, Ministry of Health and Welfare, 1-2-2 Kasumigaseki Chiyoda-Ku, 100-5 Tokyo.

Nordic: Denmark: Sundhedsstyrelsens Farmaceutiske Laboratorium, Frederiskssundsvej 378, DK-2700 Bronshoj, Denmark.

Finland: Laakintohallitus/Medicinalstyrelsen, Apteekkitoimisto/Apoteksbyran Siltasaarenkatu 18a, PL 224, SF-00531 Helsinki 53, Finland.

Iceland: Heilbrigidis-og Trygginamalaraduneytid, Lyfjanefund Laugavegur 116, 15-105 Reykjavik, Iceland.

Norway: Stabens legemiddelkontroll, Sven Oftendals vei 6, N-0950 Oslo, Norway.

Sweden: Socialsytrelsen Lakemedelsavdelning, Box 607, S-751 25 Uppsala, Sweden.

UK: Department of Health and Social Security, Market Towers, 1 Nine Elms Lane, London SW8 5NQ.

USA: Food and Drug Administration, Rockville, MD 20857, USA.

INDEX